The World in Six Songs

ALSO BY DANIEL J. LEVITIN

This Is Your Brain on Music: The Science of a Human Obsession

The World

in Six Songs

HOW THE MUSICAL BRAIN CREATED HUMAN NATURE

Daniel J. Levitin

DUTTON

DUTTON
Published by Penguin Group (USA) Inc.
375 Hudson Street, New York, New York 10014, U.S.A.
Penguin Group (Canada), 90 Eglinton Avenue East, Suite 700, Toronto, Ontario M4P 2Y3, Canada (a division of Pearson Penguin Canada Inc.); Penguin Books Ltd, 80 Strand, London WC2R 0RL, England; Penguin Ireland, 25 St Stephen's Green, Dublin 2, Ireland (a division of Penguin Books Ltd); Penguin Group (Australia), 250 Camberwell Road, Camberwell, Victoria 3124, Australia (a division of Pearson Australia Group Pty Ltd); Penguin Books India Pvt Ltd, 11 Community Centre, Panchsheel Park, New Delhi–110 017, India; Penguin Group (NZ), 67 Apollo Drive, Rosedale, North Shore 0632, New Zealand (a division of Pearson New Zealand Ltd); Penguin Books (South Africa) (Pty) Ltd, 24 Sturdee Avenue, Rosebank, Johannesburg 2196, South Africa

Penguin Books Ltd, Registered Offices: 80 Strand, London WC2R 0RL, England

Published by Dutton, a member of Penguin Group (USA) Inc.

First printing, August 2008
10 9 8 7 6 5 4 3 2 1

LIBRARY OF CONGRESS CATALOGING-IN-PUBLICATION DATA

Levitin, Daniel J.
The world in six songs: how the musical brain created human nature / Daniel J. Levitin.
p. cm.
Includes bibliographical references and index.
ISBN 978-0-525-95073-8 (hardcover)
1. Music—Psychological aspects. 2. Music—Social aspects. 3. Music—Origin. I. Title.
ML3838.L48 2008
781'.11—dc22 2008012298

Printed in the United States of America
Set in Dante MT with Agenda
Designed by Daniel Lagin

Contents

CHAPTER 1

Taking It from the Top

or "The Hills Are Alive . . ."

On my desk right now I have a stack of music CDs that couldn't be more different: an eighteenth-century opera by Marin Marais whose lyrics describe the gory details of a surgical operation; a North African griot singing a song, offered to businessmen passing by in the hopes of securing a handout; a piece written 185 years ago that requires 120 musicians to perform it properly, each of them reading a very specific and inviolable part off of a page (Beethoven's Symphony no. 9). Also in the pile: forty minutes of groans and shrieks made by humpback whales in the Pacific; a North Indian raga accompanied by electric guitar and drum machine; a Peruvian Andes vocal chorus of how to make a water jug. Would you believe an ode to the gustatory pleasures of homegrown tomatoes?

Plant 'em in the spring eat 'em in the summer
All winter without 'em's a culinary bummer

I forget all about all the sweatin' and diggin'
Every time I go out and pick me a big 'un

Homegrown tomatoes, homegrown tomatoes
What'd life be without homegrown tomatoes?
Only two things that money can't buy
That's true love and homegrown tomatoes

(Guy Clark)

That all these are music may seem self-evident to some, or the stuff of argument to others. Many of our parents or grandparents or children say that the music *we* listen to isn't music at all, it's just noise. Noise by definition is a set of sounds that are random, confused, or uninterpretable. Could it be that all sound is potentially musical if only we could understand its internal structure, its organization? This is what the composer Edgar Varèse was driving at when he famously defined music as "organized sound"—what sounds like noise to one person is music to another, and vice versa. In other words, one man's Mozart is another's Madonna, one person's Prince is another's Purcell, Parton, or Parker. Perhaps there is a key to understanding what is common to all these collections of sounds, and to what has driven humans since the beginning to engage with them so deeply as not just sound but music.

Music is characterized both by its ubiquity and its antiquity, as the musicologist David Huron notes. There is no known culture now or anytime in the past that lacks it, and some of the oldest human-made artifacts found at archaeological sites are musical instruments. Music is important in the daily lives of most people in the world, and has been throughout human history. Anyone who wants to understand human nature, the interaction between brain and culture, between evolution and society, has to take a close look at the role that music has held in the lives of humans, at the way that music and people co-evolved. Musicologists, archaeologists, and psychologists have danced around the topic, but until now, no one has brought all of these disciplines together to form a coherent account of the impact music has had on the course of our social history. This book is a lot like making a family tree, a tree of musical themes that have shaped our ancestors' lives: their working days, their sleepless nights—the soundtrack of civilization.

Anthropologists, archaeologists, biologists, and psychologists all study human origins, but relatively little attention has been paid to the origins of music. I find that odd. Americans spend more money on music than they do on prescription drugs or sex, and the average American hears more than five hours of music per day. We know now that music can affect our moods and our brain's chemistry. On a day-to-day level, a better understanding of the common history between music and humanity can help us to better understand our musical choices, our likes and dislikes, to harness the power of music to control our moods. But far more than that, understanding our mutual history will help us to see how music has been a shaping force, how music has been there to guide the development of human nature.

The World in Six Songs explains, at least in part, the evolution of music and brains over tens of thousands of years and across the six inhabited continents. Music, I argue, is not simply a distraction or a pastime, but a core element of our identity as a species, an activity that paved the way for more complex behaviors such as language, large-scale cooperative undertakings, and the passing down of important information from one generation to the next. This book explains how I came to the (some might say) radical notion that there are basically six kinds of songs that do all of this. They are songs of friendship, joy, comfort, knowledge, religion, and love.

In trying to understand the evolution of humanity and the role that music has played in it, it seems wise to begin with open minds (and ears) and not exclude any form of music too soon. However, the evolution of mind and music is easiest to follow in music that involves lyrics, because the meaning of the musical expression is less debatable. When the notes are hung on words (or is it that the words are hung on notes?), the meaning is easier to talk about usefully. Because music wasn't recorded until about a hundred years ago, nor even accurately notated until a few hundred years before that, the historic record of music is substantially lyrics. For these two reasons, music with lyrics will be the predominant focus of *The World in Six Songs.*

Much of the world's music is now available on compact disc, or on the medium that is rapidly replacing it, digitized sound files on computers (generically—and somewhat inaccurately—referred to as MP3s). We live in a time of unprecedented access to music. Virtually every song ever recorded in the history of the world is available on the Internet somewhere—for free. And although recorded music represents only a small proportion of all the music that has ever been sung, played, and heard, there is so much of it—estimates suggest 10 million songs or more—that recorded music is as good a place as any to start to talk about the music of the world. Thanks to intrepid musicologists and anthropologists, even rare, indigenous, and preindustrial music is now available to us. Cultures that have been cut off from industrialization and Western influence have had their music preserved, and by their own accounts, it may have been unchanged for many centuries, giving us a window into the music of our ancestors. The more I listen to music like this and to Western artists that are new to me, the more conscious I become of how large music is and how much there is to know.

The diversity of our musical legacy includes songs that tell stories about people, such as "Bad, Bad Leroy Brown" or "Cruella de Vil"; there's a catchy song about a murderous psychopath who kills the judge at his own trial; songs exhorting us to buy this meat product and not that (Armour hot dogs versus Oscar Mayer wieners); a song promising to keep a promise; a song mourning the loss of a parent; music made on instruments believed to be one thousand years old and on instruments invented just this week; music played on power tools; an album of Christmas carols sung by frogs; songs sung to enact social and political change; the fictional Borat singing the equally fictional national anthem of Kazakhstan, boasting about his country's mining industry:

Kazakhstan greatest country in the world
All other countries are run by little girls
Kazakhstan number one exporter of potassium
All other countries have inferior potassium

and a song about suburban noise pollution:

Here comes the dirt bike
Beware of the dirt bike . . .
Brainwashing dirt bike
Ground-shaking dirt bike
Mind-bending dirt bike
In control
Soul-crushing dirt bike

In spite of all this diversity, I have come to believe that there are basically six kinds of songs, six ways that we use music in our lives, six broad categories of music. No less.

I have been making and studying music for most of my life—I had a career producing pop and rock records for a number of years and now I direct a research laboratory studying music, evolution, and the brain. Yet I was concerned when I started this project that I might be blinkered. I didn't want to discover I was being ego- or ethnocentric. I didn't want to be culturally biased, or fall prey to any of a number of other insidious biases of gender, genre, or generation, or even pitch bias or rhythm bias. So I asked a number of musician and scientist friends what they thought all music has in common.

I visited Stanford University to meet with my old friend Jim Ferguson, who is the chairman of the Anthropology Department there; we went to high school together and have been close friends for thirty-five years. Anthropologists study culture, how it shapes our thoughts, ideas, and our worldview, and I thought for sure Jim would help me to avoid all the pitfalls and prejudices that I feared could be so seductive. Jim and I discussed how songs have many roles in the daily lives of people throughout the world and that over the millennia music has been used in so many ways we can't hope to enumerate them all.

Ubiquitous are work songs, blood songs, lust and love songs. . . . There are songs about how great God is, songs about how our god is better than yours; songs about where to find water or how to make a canoe; songs to put people to sleep and to help them stay awake. Songs with lyrics, songs of grunting and chanting, songs played on pieces of wood with holes in them, on tree trunks, with sea and turtle shells, songs made by slapping your cheeks and chest Bobby McFerrin style. I asked Jim what all these types of music had in common. His answer was that this was the *wrong question.*

Quoting the great anthropologist Clifford Geertz, Jim persuaded me that the *right* question to ask, in trying to understand music's universality, is not what all musics have in common, but how they differ. The notion that humanity can be best appreciated by extracting those features common to all cultures is a bias that I held without even knowing it. Ferguson—and Geertz—feel that the best way, perhaps the *only* way, to understand what makes us most human is to thrust ourselves face-to-face with the enormous diversity of things that humans do. It is in the particulars, the nuances, the overwhelming *variety* of ways we express ourselves that one can come to understand best what it means to be a musical human. We are a complicated, imaginative, adaptive species. How adaptable are we? Ten thousand years ago humans plus their pets and livestock accounted for about 0.1 percent of the terrestrial vertebrate biomass inhabiting the Earth; now we account for 98 percent. Humans have expanded to live in just about every climate on the surface of the Earth that is even remotely habitable. We're also a highly *variable* species. We speak thousands of different languages, have wildly different notions of religion, social order, eating habits, and marriage rites. (Kinship definitions alone account for mind-boggling variability among us, as any introductory college anthropology text will attest.)

The right question then, after due consideration of music's diversity, is whether there is a set of functions music performs in human relations. And how might these different functions of music have influenced the evolution of human emotion, reason, and spirit across distinct intellectual and cultural histories? What role did the musical brain have in shaping human nature and human culture over the past fifty thousand years or so? In short, *how did all these musics make us who we are?*

The six types of songs that shaped human nature—friendship,

joy, comfort, knowledge, religion, and love songs—I've come to think are obvious, but I accept you may take some persuading. The people of a given time or place may not have used all six. The use of some has ebbed while others flowed. In modern times with computers, PDAs, even since the beginning of written language, we haven't needed to rely so much on knowledge songs to encapsulate collective memory for us, although most English-speaking schoolchildren still learn the alphabet through song and the number line through counting songs, such as the politically incorrect "One Little Two Little Three Little Indians." For many of the world's still preliterate cultures, memory and counting songs remain essential to everyday life. As the early Greeks knew, music was a powerful way of preserving information, more effective and more efficient than simple memorizing, and we are now learning the neurobiological basis for this.

By definition, a "song" is a musical composition intended or adapted for singing. One thing the definition leaves unclear is who does the adapting. Does the adaptation have to be constructed by a professional composer or orchestrator, as when Jon Hendricks took Charlie Parker solos and added scat lyrics (nonsense syllables) to them, or when John Denver took Tchaikovsky's Fifth Symphony and added lyrics to the melody? I don't think so. If I sing the intro guitar riff to "(I Can't Get No) Satisfaction" by the Rolling Stones (as my friends and I used to do frequently when we were eleven years old), *I* am the one who has done the adapting, and even if separated from the vocal parts of that song, this melodic line then stands alone and *becomes* a "song" by virtue of my friends and I singing it. More to the point, you can sing "As Time Goes By" with the syllable "la" and never sing the words—you may have never seen *Casablanca* and you may not even *know* that the composition has words—and it becomes a song by virtue of you singing it. For that matter, suppose that only one person in the world knew the words to "As Time Goes By," and that all of us went on blissfully humming, whistling, and la-la-la-ing the melody. My intuition here is that just because we didn't sing words wouldn't mean that it wasn't a song.

Most of us share an intuition that "song" is a broad category that includes anything we might sing or any collection of sounds that resembles such a thing. Again, *The World in Six Songs* is not, I hope, culturally narrow-minded. African drum music has an important role in the daily lives of millions of people and might not strike some as being songs, but to ignore such purely *rhythmic* (and difficult to sing, unless you're Mel Tormé or Ray Stevens) forms of expression would betray a bias toward melody. The rock, pop, jazz, and hip-hop that are the most popular forms of music today would not exist without the African drumming that they evolved from. As I will show, drumming, among its many qualities, can produce powerful songs of friendship.

I have used the word *song* as a convenient shorthand and, in its most inclusive sense, as a stand-in for music in all its forms, to refer to any music that people make, with or without melody, with or without lyrics. I'm particularly interested in that portion of musical compositions that people remember, carry around in their heads long after the sound has died out, sounds that people try to repeat later in time, to play for others; the sounds that comfort them, invigorate them, and draw them closer together. I confess that I unwittingly came to this project with the bias that the best songs become popular and are sung by many. Maybe my background in the music industry put that bias in place. After all, "Happy Birthday" has been translated into nearly every language on earth (even into Klingon, as fans of *Star Trek: The Next Generation* can attest; the song is called "*qoSlIj DatIvjaj*").

Pete Seeger set me straight on this, telling me about how in some cultures, the best songs are meant to be sung and played for only one other person! Seeger is the great folk singer-songwriter who penned such songs as "Where Have All the Flowers Gone?", "If I Had a Hammer," and "Turn, Turn, Turn" (the latter with lyrics taken from Ecclesiastes).

"Among American Indians," Seeger explained, "a young man got his eye on a girl and he would make a reed flute and compose a melody. And when she came down to get a pail of water at the brook, he would hide in the weeds and play her his tune. If she liked it, she followed and saw where things led. But it was her special tune. A tune wasn't thought of as being free for everybody. It belonged to one person. You might sing somebody's song after they're dead to recall them, but each person had a private song. And of course today, many small groups feel their song belongs to them and they're not happy when it becomes something that belongs to everybody."

The fact is we are all biased to some degree by our specific life history and culture. I carry the biases of an American male growing up in California in the 1950s and 1960s. But I was lucky to have been exposed to a wide variety of music. My parents took me to see ballet and musicals before I was five, and through them (*The Nutcracker* and *Flower Drum Song*) I gained an early appreciation for Eastern scales and intervals—neuroscientists now believe that such early exposure to other tonal systems is important for later appreciation of music outside one's own culture. Just as all of us can acquire any of the world's languages as young children if we are exposed to them, so too can our brains learn to extract the rules and the structures of any of the world's musics if we're exposed to them early enough. This doesn't mean that we can't learn to speak other languages later in life, or learn to appreciate other musics, but if we encounter them as young children, we develop a natural way of processing them because our brains literally wire themselves up to the sounds of these early experiences. Through my father I developed a love of big bands and swing, through my mother a love of piano music and Broadway standards. My mother's father loved Cuban and Latin music, as well as Eastern European folk songs. Hearing Johnny Cash on the radio when I was six conditioned me for country, blues, bluegrass, and folk music.

A sentiment that I've heard many times is that classical music cannot be compared to anything else. "How can you honestly say that that repetitive, loud *garbage* called rock and roll is even *close* to the sublime *music* of the great masters?" To take this position is to ignore the inconvenient fact that a major source of joy and inspiration for the great masters themselves was the "common" popular music of their day. Mozart, Brahms, and even great-grandaddy Bach took many of their melodic ideas from ballads, bards, European folk music, and children's songs. Good melody (let alone rhythm) knows no boundaries of class, education, or upbringing.

Most of us could effortlessly construct a list of our favorite songs, of songs that just make us feel joyful, or comforted, or spiritual, that remind us of who we are, who we loved, of groups we belong to. When I ask people to do this in my laboratory, it is always surprising to see how diverse these lists are. Music is *large*. It is made by as many different types of people, with as many different backgrounds, as there are listeners. New forms of music are being invented and evolving from earlier forms every day. And each new song is a link in a millennia-long chain of evolutionary enhancements to previous song building—slight alterations in the "genetic structure" of one song lead us to a new one.

Some songs celebrate a particular individual, but then become enhanced (or diluted) by overapplication and overgeneralization.

Anyone named Maria or Michelle in the 1960s (think Bernstein and Beatles) or Alison or Sally in the 1970s (think Elvis Costello and Eric Clapton) knows what it is like to be accosted by the song bearing your name, mentioned by a friend or new acquaintance intoxicated by his own wit at having made this childishly simple connection. Anyone who has the lack of common sense to actually *sing* you the song with your name in it suffers from the doubly foolish notion that she was the first one to think of doing so. My own past has been bothered, annoyed, and taunted by endless choruses of "Danny Boy" or "Daniel" (Elton John), by people expecting me to howl at their cleverness. Steely Dan have made it a habit, a fashion even, to elevate main characters with uncommon names like Rikki, Josie, and Dupree. But the rarer the name, of course, the more exuberant is the tormentor. I have actually known people named Maggie Mae, Roxanne, Chuck E., and John-Jacob (think Rod Stewart, the Police, Rickie Lee Jones, and an old children's song), and they are astonished when people sing these songs to them as though no one has ever thought to do this before.

Friendship songs like "Smokin' in the Boy's Room" and "Tobacco Road" legitimized and banded together tens of thousands of high school (or even junior high school) students who were otherwise marginalized at the fringes of their school, engaging in an illegal but oh-it-seems-so-cool activity. School songs and national anthems are an extension of this banding together song on increasingly larger scales, the ultimate perhaps being songs uniting the entire world, such as the Michael Jackson/Lionel Richie composition "We Are the World." This sort of group formation and reinforcement finds its expressions in songs of friendship, and there is evidence that this type of song served a very important function throughout human history.

Love songs also bind people together; they express a love desired, a love found, or a love lost. They reflect a bond powerful enough to make people do things that are not always in their own best, personal interest. As Percy Sledge sang, when a man loves a woman, he'll spend his last dime trying to hold on to the woman he needs.

He'd give up all his comforts
And sleep out in the rain
If she said that's the way it ought to be.

Why does music have such power to move us? Pete Seeger says it is because of the way that medium and meaning combine in song, the combination of form and structure uniting with an emotional message.

"Musical force comes from a sense of form; whereas ordinary speech doesn't have quite that much organization. You can say what you mean, but similarly with painting or with cooking, or other arts, there is a form and design to music. And this becomes intriguing, it becomes something you can remember. Good music can leap over language barriers, and barriers of religion and politics."

The powerful mix of emotion and cultural evolution in our musical brains produced diversity, power, even history. And it has done it in six definable ways.

The study of human behavior has undergone a revolution in the past twenty years, as the methods of neuroscience have been applied to cognition and the musical experience. We can now actually see the brain at work, mapping those regions that are active during certain activities. Together with the work of evolutionary biologists, neuroscientists are beginning to build a picture of how the human brain has become adapted for thought, and to formulate theories about why it evolved the way it did. Part of my goal in writing this book is to bring these perspectives to bear on the question of music, the brain, culture, and thought. If music has lasted so long in our species, what are the cultural and biological forces driving its forms and uses?

In the beginning, there was no language. We may have had music before we had a word for it. We had sounds, of course, and they communicated to us. Thunder, rain, and wind. The sound of boulders or avalanches rolling down hills. The warning calls of birds and monkeys. The growls of lions and tigers and bears (oh, my!). And sights and smells added to our awareness of things happening-in-the-world, sometimes benign, sometimes a warning. Before language, we were critically limited in our ability to represent what wasn't there. Was this a limitation of our brains or simply that we lacked verbal communication, words to serve as placeholders for things that weren't immediately in our consciousness?

The evidence from evolutionary neuroscience suggests that these are really the same question. We usually think of evolution as governing our physical bodies—the opposable thumb, walking upright, depth vision—but our brains evolved as well. Before there was language, our brains did not have the full capacity to learn language, to speak or to represent it. As our brains developed both the physiological and cognitive flexibility to manipulate symbols, language emerged gradually, and the use of rudimentary verbalizations—grunts, calls, shrieks, and groans—further stimulated the growth potential for the types of neural structures that would support language in the broadest sense. So how did language and music happen—who invented them and where did they come from?

It is unlikely that either language or music was invented by a single innovator or at a single place and time; rather, they were shaped by a large number of refinements, contributed to by legions of developers over many millenia and throughout all parts of the world. And they were no doubt crafted upon structures and abilities that we already had, structures we inherited genetically from protohumans and our nonhuman animal ancestors. It's true that human language is qualitatively different from any animal language, specifically in that it is generative (able to combine elements to create an unlimited number of utterances) and self-referential (able to use language to talk about language). I believe that the evolution of a single brain mechanism—probably located in the prefrontal cortex—created a common mode of thought that underlies the development both of language and of art.

This new neural mechanism gave us the three cognitive abilities that characterize the musical brain. The first is *perspective-taking*: the ability to think about our own thoughts and to realize that other people may have thoughts or beliefs that differ from our own. The second is *representation*: the ability to think about things that aren't right-there-in-front-of-us. The third is *rearrangement*: the ability to combine, recombine, and impose hierarchical order on elements in the world. The combination of these three faculties gave early humans the ability to create their own depictions of the world—paintings, drawings, and sculpture—that preserved the essential features of things though not necessarily the distracting details. These three abilities, alone and in combination, are the common foundation of language and art. Language and art both serve to represent the world to us in ways that are not exactly the world itself, but which allow us to preserve essential features of the world in our own minds, and to convey what *our* minds perceive to others. The awareness that what *we* are feeling is not necessarily what another is feeling, coupled with our drive to create social bonds with others, gave rise to language and art, to poetry, drawing, dance, sculpture . . . and music.

An important property of language is that we can talk about things that are not there. We can talk *about* fear without actually being scared, or talk about the word *fear* without having any feelings of fright. All this representing requires massive computational power. To support this kind of abstract thought, our brains had to evolve to handle billions of bits of simultaneous, often contradictory, information and to connect those bits to other things that came before and will come again.

One of the things that humans are good at and animals are not is encoding *relations*. We can easily learn the idea of one thing being bigger than another. If I ask even a five-year-old to select the *largest* of three blocks in front of her, she will do this effortlessly. If I then bring in a new block that is twice the size of the one she just selected, she can shift her thinking, and choose *that* when I re-ask the question. A five-year-old understands this. No dog can do this, and only some primates.

This understanding of *relations* turns out to be fundamental for music appreciation; it is a cornerstone of all human musical systems. One such musical relation is *octave equivalence*, the principle that allows men and women to sing a song together and sound like they're singing in unison, even though the women are (typically) an octave higher. This relative mode of processing also permits us to recognize "Happy Birthday" as the same song regardless of what key it is sung in, a process musicians call *transposition.* It is also the basis of composition in nearly every musical style we know of. Take the opening to Beethoven's Fifth for example. We hear three notes of the same pitch and duration, followed by a longer note at a lower pitch. Beethoven takes this pattern and moves it lower in the scale, so that the next four notes follow the same contour and rhythm. Our ability to recognize that this pattern is essentially the same, even though none of the notes are the same, is relational processing. Decades of research on music cognition have shown that humans process music using both absolute and relational processing—that is, we attend to the actual pitches and duration we hear in music, as well as their relative values. This dual mode of processing is rare among species, having not yet been confirmed in any species other than our own.

These modes of processing and the brain mechanisms that gave rise to them were necessary for the development of language, music, poetry, and art. And as I said earlier, I believe they were made possible by the evolution of a common brain structure. All art seeks to represent some aspect of human experience, and it does so selectively. If an artistic object represents the thing-itself perfectly, it is just another copy of that thing. The point of art is to emphasize some elements at the expense of others—to focus on one or more aspects of the thing's visual or auditory appearance or of the way we feel about it—in order to call particular attention to them. We may do it so that we can remind ourselves of how we felt about a certain experience, or to communicate that experience to others. Music combines the temporal aspects of film and dance with the spatial aspects of painting and sculpture, where pitch space (or frequency space) takes the place of three-dimensional physical space in the visual arts.

The brain has even developed frequency maps in the auditory cortex that function much the way that spatial maps do in the visual cortex.

Our drive to create art is so powerful that we find ways to do it under the greatest hardships. In the concentration camps of Germany during World War II, many prisoners spontaneously wrote poetry, composed songs, and painted—activities that, according to Viktor Frankl—gave meaning to the lives of those miserably interred there. Frankl and others have noted that such creativity under exceptional circumstances is not typically the result of a conscious decision on the part of a person to improve his outlook or his life through art. To the contrary, it presents itself as an almost biological need, as essential a drive as that for eating and sleeping—indeed, many artists, absorbed in their work, temporarily forget all about eating and sleeping.

Ursula Bellugi of the Salk Institute discovered a form of poetry invented by people who are deaf and who communicate using American Sign Language. Rather than using a single hand to make certain signs, they'll use two—holding one sign in the air with the left hand while the right hand takes over, creating a legato, overlapping visual. The signs will be altered, creating certain visual repetitions—a visual music—that are analogous to verbal repetitions, phrasing, and meter in spoken poetry. We create because we cannot stop ourselves from doing so. Because our brains were made that way. Because evolution and natural selection favored those brains that had a creative impulse, one that could be turned toward the service of finding shelter or food when others were unable to find it; toward enticing mates to procreate and care for children amid competition for mates. Creative brains indicated cognitive and emotional flexibility, the kind that could come in useful on the hunt or during interpersonal or intertribal conflicts.

Creative brains became more attractive during centuries of sexual selection because they could solve a wider range of unanticipatable problems. But how did musical brains become attractive? Consider why we find babies attractive as an analogy. Suppose that some people, due to random processes that we don't understand (and that they don't understand either!), happen to find babies cute and other people do not. These random processes are formally similar to the ones that make you taller than your father, make you go bald at twenty-five, give you a keen sense of direction, or the ability to laugh when all around you is falling apart. The people who find babies cute—again, they may have simply won some sort of genetic lottery and be the first ones in their family to have this characteristic—are going to spend a lot more time with their babies, nurturing them, playing with them, and attending to them, compared to the people who just don't find babies all that cute. Over millions of cases like this, the parents who just happen to find babies cute will tend to have babies that grow up to be more well adjusted, well educated, and healthy than the other parents' babies. This difference in upbringing may well cause the nurtured babies to be more likely to find mates and have babies of their own, if only because they were more likely to be healthy and live that long, or to have the knowledge and support system necessary to acquire food and shelter when it is their time to mate. In the long run, the offspring of these nurtured babies will tend, therefore, to outnumber the offspring of the non-nurtured babies.

This is the basic principle of Darwinian natural selection. In other words, as the philosopher Daniel Dennett points out, we don't think babies are cute because they are intrinsically or objectively cute (whatever that would mean). Rather, the process of evolution favored those people and their offspring who found babies cute, and in turn this characteristic became widely distributed in the population.

By analogy then, humans who just happened to find creativity attractive may have hitched their reproductive wagons to musicians and artists, and—unbeknownst to them at the time—conferred a survival advantage on their offspring. Early musicians may have been able to forge closer bonds with those around them; they may have been better able to communicate emotionally, diffuse confrontation, and ease interpersonal tensions. They may also have been able to encode important survival information in songs, an easily memorable format that gave their children an additional survival advantage. Making and listening to music, then, feels good not because of anything intrinsic in the music. Rather, those of our ancestors who just happened to feel good during musical activities are the ones who survived to pass on the gene that gave rise to these feelings.

One important thing that makes us human, one thing we have that separates us from all other species on our planet, has been noted by psychologists and biologists. It's not the fact that we have a language to communicate with—other animals, such as birds, whales, dolphins, even bees, have sophisticated signaling systems. It's not that we've learned to use tools (chimpanzees do that), that we have built societies (ants have those), or learned to deceive (crows and monkeys). It's not that we're bipedal and have opposable thumbs (primates) or that we often mate for life (gibbons, prairie voles, angelfish, sandhill cranes, termites). What distinguishes us most is one thing no other animals do: *art*. And it's not just the existence of art, but the centrality of it. Humans have demonstrated a powerful drive toward making art of all different kinds—representational and abstract, static and dynamic, creations that employ space, time, sight, sound, and movement.

Our urge toward artistic expression shows up in cave paintings, and decorations on otherwise solely utilitarian items, such as thirty-thousand-year-old water jugs. Some of the earliest cave paintings show humans dancing. Nearly one hundred years ago the *Encyclopedia Britannica,* in its 1911 edition, stated that poetry had exerted "as much an effect upon human destiny as . . . the discovery of the use of fire." Equating poetry with fire is both metaphorically satisfying and dramatic (the fire in men's and women's souls? the burning desire to express feelings with rhythm and rhyme?). But are we meant to believe that poetry actually exerted such a profound effect on the course of human events? *Britannica* argues just this—that poetry, and presumably lyrics, have changed history, started and stopped wars, documented the history of humankind, and changed men's [sic] minds about the course of their lives.

Apart from signaling creativity and the ability to engage in abstract thinking, the development of the artistic (poetic, musical, dancing, and painting) brain allowed for the metaphorical communication of passion and emotion. Metaphor allows us to explain things to people in indirect ways, sometimes avoiding confrontation, sometimes helping another to see that which she has difficulty understanding. Art allows us to focus another's attention on aspects of a feeling or a perception that he might not otherwise see, literally framing the point of interest in a way that it becomes separated from a background of competing ideas or perceptions.

The auditory arts of music and poetry hold a privileged position in human history, and we see this reflected in our own time in neurological case studies. Individuals suffering from Alzheimer's disease, victims of strokes, tumors, or other organic brain trauma, may lose the ability to recognize faces, even of people they've known their entire lives. They may lose the ability to recognize simple objects such as hairbrushes or forks. But many of these same patients can still recite poetry by heart, and sing songs that they knew as children. Verse—whether spoken or sung—appears to be deeply encoded in the human brain. Many artists throughout history have felt an overwhelming drive to write music and poetry, on battlefields, in dungeons, on their deathbeds. This drive no doubt arose from those same frontal cortex mutations and adaptations that made art possible in the first place, the structural changes that gave rise to language and art in general. We write and recite music and poetry not because it feels good intrinsically, but because those ancestors of ours for whom it felt good are the ones who survived and reproduced, passing on this visceral preference. We are a musical species today because our ancestors were, going back tens of thousands of years.

But we are not today as much of a poetic species as the historical record suggests we used to be, having traded poems for songs over the last several hundred years. The average fourteen-year-old will hear more music in a month than my grandfather heard in his entire life. An iPod today easily holds twenty thousand songs, more than the libraries of seven urban radio stations, an order of magnitude more songs than an entire tribe of our hunter-gatherer ancestors would have encountered in their entire lives. Before looking at the six songs that shaped human nature, it is important to take a closer look at just what a song is and isn't with respect to its lyrics; whether the words are poetry by definition (or what poetry actually is, if indeed it is something different). Do lyrics convey the same meaning when they are divorced from their music?

In the introduction to his book of lyrics, Sting writes:

> The two, lyrics and music, have always been mutually dependent, in much the same way as a mannequin and a set of clothes are dependent on each other; separate them, and what remains is a naked dummy and a pile of cloth. Publishing my lyrics . . . [invites] the question as to whether song lyrics are in fact poetry or something else entirely. . . . My wares have . . . been shorn of the very garments that gave them their shape in the first place.

The shape of lyrics is influenced by different things than the shape of poetry—the melody and rhythms of music provide an extrinsic framework, whereas poetry's structure in intrinsic. In music, some notes are accented relative to others, by virtue of their pitch, loudness, or rhythm; these accents constrain the words that will fit well with the melody, and help establish the musical mannequin on which the lyrical clothing will be hung. In poetry, on the other hand, different conventional structures and forms that poets impose on themselves carry meaning—epics, elegies, and odes may signal history, mourning, and love. Traditionally poetry has been discussed in terms of these forms (rhyming patterns, metrical patterns, number of lines). Sonnets were for love. Epics suited couplets. Dirges were for misery.

I've been writing songs all my life, but my friends who are *real* songwriters tell me that while my melodies are strong, my lyrics are not. I think of myself as a verbal person, so the irony of this is clear to me. I have to confess that for most of my life I never engaged much with poetry, and didn't really understand it, in spite of having taken a course in it in college in the 1970s. About ten years ago, my friend Michael Brook (a composer of instrumentals and film scores) suggested that if I wanted to make my lyric writing better, I should read poetry.

The next day, coincidentally, I ran into my old poetry professor, I'll call him Lee, on campus where—following a most unlikely path through the music industry—I had become a professor myself. We went for coffee together and I told him of my desire to become a better lyricist. I asked him to explain how poetry and song lyrics differ. Once again, I found I was asking the wrong question!

"Lyrics *are* poetry," Lee explained. "They are two varieties of the *same thing.* The lyrics of popular songs are only a particular kind of poetry. You seem to believe there exists an absolute distinction between the two, and there doesn't. Lyric poetry has been around since the beginnings." Lee mentioned the treasure trove of medieval and Elizabethan lyrics—the poems/songs of Campion, Sydney, Shakespeare (and Schubert's "Who Is Sylvia?", for example, and other lieder). He pointed out that it is not uncommon for written poetry to be later rendered into song, as in Jonson's "Drink to Me Only with Thine Eyes," or Burns's poems or William Bolcom's music-making with Blake and Roethke.

This sentiment was echoed more recently by John Barr, president of the Poetry Foundation, and himself a respected poet. In defending poetry not of the ivory tower sort, he writes:

> People who care about their poetry often experience genuine feelings of embarrassment, even revulsion, when confronted with cowboy poetry, rap and hip-hop, and children's poetry. . . . Their readerly sensibilities are offended. (If the writing gives them any pleasure, it is a guilty pleasure.) The fact that Wallace McRae, Tupac Shakur, and Jack Prelutsky wrote these works for large, devoted audiences simply adds insult to the injury. Somewhat defensively, the serious poetry crowd dismisses such work as verse, not poetry, and

> generally acts so as to avoid it, if at all possible, in the future. . . . The result is a poetry world of broad divides.

Of course song lyrics do have something that conventional poems don't—the melody, the mannequin on which Sting was saying he had hung his lyrical clothes. That is, most poems, by definition, have to convey an emotional message through some combination of rhyme, meter (the way the sounds are organized in time, including their accent structure), metaphor, and verbal imagery that add up to great beauty of expression. They also must convey a sense of movement—a forward, rhythmic momentum all their own. Song lyrics may do all these things, but they don't have to. They always have the music there to help them along, melodies and harmonies that can provide accent structures, forward motion, and a kind of harmonic-textual context. In other words, lyrics are not *intended* to stand alone (and as to the words of poems, to quote another Sting lyric, "they dance alone").

Lee and I met once a week for that academic term. He brought in some of his favorite written poetry; I brought in my favorite popular music lyrics (which he never failed to remind me were also poetry). I came to see that, whatever its form, written poetry is characterized by a kind of music. Accent structures in words naturally make a sort of melody. In the word *melody* itself the first syllable is stressed, which makes it louder than the others, and most native English speakers will give it a higher pitch than the other syllables. The word *melody* has a melody! Good poetry plays with speech sounds to create a pleasing set of pitch patterns, and good poetry contains rhythmic groupings that are songlike. When a poem succeeds, it is a sensual experience—the way the words feel in the mouth of the speaker and the way they sound in the ears of the hearer are part of the encounter. Unlike prose, most poems ask to be read aloud. This is why poetry lovers usually do so. Just reading the poem is not enough. The reader needs to *feel* the rhythms. Song lyrics ask to be sung; reading them doesn't typically convey all the nuance of expression that was imbued in them during their creation.

Occasionally, a song lyric *can* stand completely on its own, but Lee was quick to point out that this doesn't make it any better than one that cannot; it is simply another feature of that particular piece of writing. But through our weekly meetings, I gained a deep appreciation for the interplay between sound and form, between meaning and structure, that characterizes both forms of writing.

One characteristic of poetry and lyrics, compared to ordinary speech or writing, is compression of meaning. Meaning tends to be densely packed, conveyed in fewer words than we would normally use in conversation or prose. The compression of meaning invites us to interpret, to be participants in the unfolding of the story. The best poetry—the best art in any medium—is ambiguous. Ambiguity begets participation. Poetry slows us down from the way we normally use language; we read and hear poetry and stop thinking about language the way we normally do; we slow down in order to contemplate all the different reverberations of meaning it contains.

The spiritual or emotional aspects of art are perhaps their most important qualities. Poetry is no exception—it is written in order to capture feelings and personal, subjective interpretations of events, rather than to deliver a mere description—you might say that it is the right-brain equivalent of a news report. As Helen Vendler (a Harvard professor and leading poetry critic) says, "Poems are hypothetical sites of speculation, not position papers.

They do not exist on the same plane as actual life; they are not votes, they are not uttered from a podium or pulpit, they are not essays. They are products of reverie."

Once in a while we run into people who can recite poetry from memory. We all know people who memorize song lyrics and drop them into conversation at opportune moments. What makes a good lyric or poem? That it is easy to remember? I have lyrics bouncing around in my head all the time, and they are released from their neural prison at even the slightest provocation. During an uncharacteristic weeklong rainstorm at Stanford, it seemed as though my brain had a mind of its own (!), calling up one rain song after another. It began in one of those Jungian synchronicity experiences that Sting writes about. I was listening to the song "Rain" when I heard a crack of thunder followed by a few taps of light droplets on my roof. Within minutes, the rain was pounding. I raced outside to put the top up on my car (California—it was a convertible of course) and to bring in the dog, who was already cowering underneath the hydrangeas. I had the first verse of that Beatles song stuck in my head ("When the rain comes/they run and hide their heads"), and to get it unstuck I tried to think of another song. The first one that came to mind was "Raindrops Keep Fallin' on My Head" by Bacharach and David, a great song, but one that—from hard-won experience—I knew would be stuck in my head for a solid week if I didn't nip this one in the bud, and fast.

It's funny how memory works—instead of a segue to another Bacharach and David composition ("Do You Know the Way to San Jose?" or "I Say a Little Prayer") or another B. J. Thomas record ("Hooked on a Feeling," "I Just Can't Help Believing") or another song with the same I-I maj 7-I7-IV chord progression ("Everybody's Talkin'," "Something"), my frontal cortex was bent on searching my hippocampus for a song with "rain" in the title, and instantaneously, unconsciously, my brain delivered "You and Me and Rain on the Roof" by the Lovin' Spoonful. I love this song. The melody descends the scale from the fifth degree, *sol,* down an octave to the *sol* below, suggesting the Greek lydian mode, and I knew from experience that there was little danger this would get stuck in my head, but it would at least push out the Bacharach melody. Why was I trying to get rid of that? Oh yes, because the Beatles' "Rain" was reverberating around in there. Oh no! Soon I had that back. Quick! Think of John Sebastian. Ahhhh. "You and me and rain on the roof . . ." Sol - fa - mi - re - do - sol - la - sol.

It rained all day. Puddles started to gather and the little sewer drain on every corner started to back up. Water began to gather in street intersections. The city engineers had not had to design for water runoff because it usually doesn't rain much in this part of the world. It continued to rain for a week. My overactive hippocampus kept offering me more rain songs, floating up from my unconscious: "No Rain" by Blind Melon, "Fire and Rain" by James Taylor (and the haunting cover version by Blood, Sweat & Tears), "I Can't Stand the Rain" by Tina Turner, "Still Raining, Still Dreaming" by Hendrix, and of course "The Rain Song" by Led Zeppelin, the opening chord a downward arpeggio, itself falling like rain. I congratulated myself on successfully avoiding getting stuck with an earworm from "Here Comes the Rain Again" by Eurythmics or "Walk Between the Raindrops" by Donald Fagen. I fired up the stereo with "Rainy Days and Mondays" (the Carpenters), "Rainin'" (Rosanne Cash), "Let It Rain" (Eric Clapton with his group Derek and the Dominos), and two rain songs by one of my favorite groups: "Who'll Stop the Rain" and "Have You Ever Seen the Rain?" (Creedence Clearwater Revival). Two more of my favorite groups finally weighed in from down below in my hippocampus, playing in my head as if a CD player were wired directly to my neurons: "Prayers for Rain" by the Cure, and "Bangkok Rain" by the Cult. So many rain songs! And it kept raining outside.

When I talked to one of my favorite songwriters, Rodney Crowell, about *Six Songs*, he argued that the first songs composed by humans probably dealt with the elements, with weather, sun, moon, rain, and so on, because these would have been so central to early man.

Lee and I met the following week and the sun had been out for a couple of days by then. ("Here Comes the Sun," I thought as I walked across campus to meet him, and this gave way to auditory images of "Sun King" and "I'll Follow the Sun" [Beatles], "Let the Sunshine In" [The 5th Dimension], "Sunny" [Bobby Hebb], "You Are My Sunshine" [as performed by Ray Charles], "Wake Up Sunshine" [Chicago], "Who Loves the Sun" [The Velvet Underground], "California Sun" [the Ramones and the Dictators], "House of the Rising Sun" [Eric Burdon back when he was with the Animals, one of the first rock songs I ever wanted to play nonstop for a week].) Lee brought Robert Frost's "The Wind and the Rain" and Walt Whitman's "Give Me the Splendid Silent Sun" from *Leaves of Grass*. I brought Cole Porter and Joni Mitchell.

Many of my favorite lyrics have internal rhymes. By "internal rhymes," I'm referring to rhymes and near rhymes that occur anywhere other than the end of a line, like these from Cole Porter:

Oh by Jove and by Jehovah, you have set my heart aflame,
(And to you, you Casanova, my reactions are the same.)
I would sing thee tender verses but the flair, alas, I lack.
(Oh go on, try to versify and I'll versify back.)

Notice in the first two lines the long o sound that is repeated in Jove, JeHOvah and CasaNOva, and the near rhyme of *heart* and *are* near the end of those two lines. Another thing Porter is famous for is invoking common, everyday expressions in playful ways. We are familiar with the phrase "alas and alack," which the composer plays with when he writes, in the third line, "alas, I lack." All this while maintaining the end-of-the-line rhymes we've come to expect in contemporary song: aflame/same and lack/back.

Or consider these lines, from "Begin the Beguine," where the song title itself is a visual and auditory wordplay:

To live it again is past all endeavor,
Except when that tune clutches my heart,
And there we are, swearing to love forever,
And promising never, never to part.

Notice in the third line the internal rhyme in *there* and *swear.* The first and third lines rhyme, as they ought to, ending with *endeavor* and *forever.* But Porter adds an additional rhyme to these in the middle of the fourth line with the repetition of *never.* I sure wish I could write like that! (I'd lure bright fish, I'd swish as I sat, my heart would go pitter pat, if only I could dish out fine lines such as that!)

Of course some people don't care about this sort of wordplay as much as they do about the content; they may study lyrics intently, looking for wisdom, sage advice, just as many of us did in the sixties. Rock stars were our poets; we felt that they had hard-won life lessons to pass on to the rest of us.

Others learn the lyrics syllable-by-syllable as a means of recalling the music, but don't pay much attention to the lyrical content itself. I had a girlfriend who was born and raised in Belgium, and we spent many wonderful vacations visiting her family and friends in her hometown Mons (*Bergen* in Flemish), where she went to university at the Faculté de Polytechnique there. Every one of her friends knew the Eagles' song "Hotel California" syllable-for-syllable, but most of them didn't speak a word of English. They had no idea what they were saying when they sang "warm smell of co-li-tas/rising up through the air/up ahead in the distance/I saw a shim-mer-ing light." Not knowing English, they didn't know where the beginnings and endings of words were. Just as my little sister used to think that "The Star-Spangled Banner" spoke of a particular kind of lamp called a donzerlee light (for "dawn's early light"), my Belgian friends thought there must be a type of lamp called a "murring light" (from shim-mer-ing light). And what was this thing called a "prizzonerzeer" that all of us are ("we are all just prisoners here . . .")? They were even more curious to know what the song *meant,* and I had to confess that as much as I loved the song—I had even learned the guitar solo note-for-note to impress fellow musicians—I didn't have the slightest idea what it was about. The emotional impact of the line "you can check out anytime you like/but you can never leave" was not diminished at all by the fact I didn't know what Don Henley was trying to say.

This is the power of the song lyric—the mutually supporting forces that bind rhythm, melody, harmony, timbre, lyrics, and meaning in a song allow some of the elements to fill in for others when there is ambiguity, contradiction, or outright opacity, as is the case with "Hotel California." That the literal meaning is not apparent in that song—or for that matter, in almost any song by Steely Dan, the kings of cryptic lyrics—doesn't reduce the power of the song. Each song's elements add up to an artistic result. The whole invokes meaning but does not constrain it. In fact, this is one of the features that gives songs their power over us: because the meaning is not perfectly defined, each of us as listeners becomes a participant in the ongoing process of understanding the song. The song is personal because we've been asked or forced to fill in some of the meaning for ourselves.

Many of us feel a peculiarly intimate relationship with popular songwriters because it is their very voices that we hear in our heads. (And it is for this reason that poetry fans so highly value recordings of a favorite poet reading his or her own works.) Most of us listen to songs we like *hundreds* of times. The voice, nuances, and singer's phrasing become embedded in our memories in a way we don't get with poetry that we read to ourselves. We feel we know something about the lives, the thoughts and feelings of our favorite songwriters because we know several or dozens of their songs. And because of the mutually reinforcing constraints of rhythm, melody, and accent structure—combined with a shot of dopamine or other neurochemicals that are known to accompany music listening—our relationship with song becomes vivid and long-lasting, activating more regions of the brain than anything else we know of. The connection to some songs is so long-lasting that patients with Alzheimer's disease remember songs and song lyrics long after they've forgotten everything else.

The Beatles ushered in an era of singers writing their own songs. Although Chuck Berry wrote his, and Elvis cowrote a few of his own, it wasn't until the Beatles and their enormous commercial success—followed by the success and writing of Bob Dylan and the Beach Boys, among others—that fans began to expect musicians to write their own material. The Beatles even cultivated this sort of personal connection to their audience. In their early songs, Paul McCartney says, he and John intentionally—somewhat calculatingly—tried to inject personal pronouns into as many of the early lyrics and song titles as they could. They took seriously the task of forging a relationship with their fans in a very personal way. "She Loves *You*," "*I* Want to Hold *Your* Hand," "P.S. *I* Love *You*," "Love *Me* Do," "Please Please *Me*," "From *Me* to *You*."

Still, it is important to note that some people ignore the lyrics more or less, and are drawn primarily to rhythm and melody. Although many people are attracted by the storylines of opera, equal numbers report that they don't even try to follow the plot, enjoying simply the colorful scenery and the beautiful vocal sounds they hear. Even in pop, jazz, hip-hop, and rock, legions of people believe that the lyrics function primarily as an afterthought, something to hang the melody on. "What do lyrics have to do with music?" many demand. "They're just there so that the singer doesn't have to go 'la la la' with the melody all the time." And as far as many people are concerned, "la la la" would be just fine.

But for those who love lyrics, for whom "la la la" won't do, there are many rewards in studying the ways in which the best of them are crafted. While researching this book, Sting and I discussed the relationship between poetry and lyrics. Both of us being Joni Mitchell fans, we discussed her song "Amelia" as an example of a lyric we admire:

I was driving across the burning desert
When I spotted six jet planes
Leaving six white vapor trails
Across the bleak terrain
It was the hexagram of the heavens
It was the strings of my guitar
Oh Amelia, it was just a false alarm.

Note the repetition of the long i sound in *I* and *driving* in the first line; the repetition of the d sound in *driving* and *desert* in that same line; the repetition of the s sound in *spotted* and *six* in the second line. Of course there is also the alliteration in *hexagram* of the *heavens*. The song features a prominent guitar, connecting the music to the lyric. I love that she mentions her six-string guitar in the sixth line of the song, just one subtle element among many that create an internal consistency in this lyric. There is the semantic connection between a *desert* and a *plain*, both flat expanses of terrain, a connection implied by her choice of the homonym *planes* in the second line.

Of course, some of these connections may be only coincidence, things the writer herself did not notice. But these sorts of connections are prevalent in all great poetry, displaying the subtle workings of and intricate connections among imagination, intellect, and the subconscious. Even if a poet wasn't aware herself of all that could be read into a particular poem, great poems reward this sort of analysis, and lesser poems do not—the deeper you look, the less interesting they seem. And the imagery is palpable—a burning desert, white vapor trails, bleak terrain. The song draws pictures with words. It also has metaphor, the drive across the desert being a Lakoffian metaphor for a relationship.

Many of Sting's own lyrics have a literary sensibility, coupled with a real ease of expression—the very sensual quality I spoke of earlier. Take his song "Russians" for example:

In Europe and America
There's a growing feeling of hysteria
Conditioned to respond to all the threats
In the rhetorical speeches of the Soviets
Mister Khrushchev said, "We will bury you"
I don't subscribe to this point of view
It would be such an ignorant thing to do
If the Russians love their children too

How can I save my little boy
From Oppenheimer's deadly toy
There is no monopoly of common sense
On either side of the political fence
We share the same biology
Regardless of ideology
Believe me when I say to you
I hope the Russians love their children too

The lyrics roll right off the tongue, easily. They're easy to say, and they feel good in the mouth. Repetitions of vowel and consonant sounds—the phonology—give the verse forward momentum. The meaning is artfully veiled in metaphors. The last line of the first verse mentions children, the first line of the next verse a "boy," and then the atom bomb is described in terms of children and boys, "Oppenheimer's deadly toy." The poet delights in stringing together familiar phrases that reverberate in our collective memory—"rhetorical speeches," "we will bury you," "the political fence," and so on. The message is cast in terms of a hope that the "monsters" that inhabited each opposing side of the Cold War (for that is how we were raised to see our enemies, as subhuman monsters) will find common ground and hopefully common sense in their love for their children. This echoes General William Westmoreland's Vietnam War–era pronouncement (made famous in the chilling documentary *Hearts and Minds*) that there was no shame in accidentally killing North Vietnamese children because "the Oriental mind doesn't put the same high price on life as does the Westerner."

As we saw that good poetry must, Sting's words create a rhythmic pulse. We can see this visually by adding diacritical marks showing the accent structure. The first line begins somewhat leisurely with eight syllables, and only two of them are stressed. The second line picks up the pace with eleven syllables, three of which are stressed, and more than half of which (six) begin with consonants. This trend continues in line three, where nine of the ten syllables begin with consonants. The combined effect of all these consonants is like a series of small explosions (they literally *are* explosions as air is thrust out of your mouth, something it doesn't do with vowels) and these serve to propel the lyric forward.

"In Europe and America
There's a growing feeling of hysteria
Conditioned to respond to all the threats
In the rhetorical speeches of the Soviets
Mister Khrushchev said, "We will bury you"
I don't subscribe to this point of view
It'd be such an ignorant thing to do
If the Russians love their children too

In normal English speech, we tend to raise the pitch of syllables that are stressed or accented and lower the pitch of unstressed syllables, as is the case in many (but by no means all) languages. If we violate this in English, it becomes confounded with the rising intonation we normally use to indicate a question. For example, we would normally say the word *Eu*-rope by making the first syllable a little louder than the second and by dropping the pitch of the second (unstressed) syllable. Lowering a pitch like this usually makes the syllable sound unstressed. If instead, holding loudness manipulations the same (that is, keeping

/Eu/ no louder than /rope/), I *raise* the pitch of the second syllable, it sounds like I'm asking a question, or like I'm unsure that I've chosen the right word. (Or like I'm fifteen? Ya know? Like, where every statement sounds like it's a question? Even assertions? Like this?)

In "Russians," Sting artfully interposes pitch accents and linguistic accents. This breathes life into the lyrics by introducing the unexpected, and allowing the text and melody to mutually support (but not entirely determine) one another. Where the melody rises, it sometimes rises on syllables that are unstressed. Such a technique would not work well in a dance or funk song, where the linguistic and melodic accents need to line up in order to give an unambiguous sense of the beat. Think "I Got You (I Feel Good)" by James Brown:

I feel good, I knew that I would
. . .
I feel nice, like sugar and spice
So good, so nice, I got you

Apart from the fact that all (but one) of the words are monosyllabic, the accent structure of the melody supports the accent structure of how this might be spoken—contributing to the pounding insistence of the groove.

"Russians" delivers as a song lyric because it marries text to melody, and because the lyrics feel effortless. It succeeds as a poem because even without the melody, it conveys its own rhythm, the forward momentum created by its accent structure and use of plosive consonants.

"Amelia" and "Russians" demonstrate great beauty in language and expression, used to convey an intensely imaginative interpretation of their subjects. Rather than delivering a literal description, they effectively capture feelings and impressions of events by telling us the most evocative parts of the story—often with figurative language, instead of the sort of objective details we would get in a newspaper article. We sense in them also a drive toward art—an unstoppable internal force that impelled the writer to write. In these lyrics, as in many great works of art, we feel an inevitability about them—that they have always existed and were just waiting to be discovered. When attached to the song, the words evoke additional emotions because of the harmonic tension that the musical notes add. Together, the lyrics plus melody, harmony, and rhythm bring nuances and shades of meaning that the words alone can't deliver.

Both poetry and lyrics and all the visual arts draw their power from their ability to express abstractions of reality. When the poet Herbert Read wrote,

> Art, at the dawn of human culture, was a key to survival, a sharpening of the faculties essential to the struggle for existence. Art, in my opinion, has remained a key to survival.

I believe he was referring to this abstraction process that is intrinsic in the creation and appreciation of all artistic objects, and that is a feature of the musical brain. Drawings, paintings, sculpture, poems, and song allow the creator to represent an object in its absence, to experiment with different interpretations of it, and thus—at least in fantasy—to exert power over it. Songs and poems derive their ultimate power in this way.

Songs give us a multilayered, multidimensional context, in the form of harmony, melody, and timbre. We can experience them in many different modes of enjoyment—as background music, as aesthetic objets d'art independent of their meaning, as music to sing with friends or sing along with in the shower or car; they can alter our moods and minds. Each of the elements of melody, rhythm, timbre, meter, contour, and words can be appreciated alone or in combination. "I Got You (I Feel Good)" may not have changed the course of human history, but it has been enjoyed by millions of people over many millions of hours. To the extent that we are the sum total of all our life's experiences, it has become a part of our thoughts, and (as neuroscientists know) that means a part of the very wiring of our brains.

But that is not the same as guiding human destiny. *The World in Six Songs* is the story of just how music has changed the course of human civilization, in fact, the story of how it made societies and civilizations possible. Other art forms—poetry, sculpture, literature, film, and painting—can also fit into these functional categories, but this is the story of music and its primacy in shaping human nature. Through a process of co-evolution of brains and music, through the structures throughout our cortex and neocortex, from our brain stem to the prefrontal cortex, from the limbic system to the cerebellum, music uniquely insinuates itself into our heads. It does this in six distinctive ways, each of them with its own evolutionary basis.

I attended the annual meeting of Kindermusik teachers this summer. Parents, children, and teachers came from more than sixty countries to participate in workshops and listen to lectures. The highlight of the conference for me was the music before the keynote speech. Fifty young children, between the ages of four and twelve, sang this song, based on a traditional German folk song, accompanied with syncopated clapping and synchronized movement:

All things shall perish from under the sky
Music alone shall live
Music alone shall live
Music alone shall live
Never to die.

Pairs of children from different countries took turns at the microphone, singing lines from the song in their native languages: Cantonese, Japanese, Romanian, !Xotha, Portuguese, Arabic, with each stanza ending in the English refrain in three-part harmony:

"Music alone shall live, never to die."

And the music that will never die has been with humans since we first *became* humans. It has shaped the world through six kinds of songs: friendship, joy, comfort, knowledge, religion, and love.

CHAPTER 2

Friendship

or "War (What Is It Good For?)"

I. It is the near twilight hour, the fog hanging thick, close to the ground like a heavy, leaden weight. Imagine that you're an early human, sleeping with your villagers, huddled together on the ground, near the dying embers of the fire circle. First there is a feeling that disturbs your sleep—not all of your group is roused, but you notice that a few others were also disturbed—by a vibration more than a sound. And you wonder: Was it real or a dream? Rumbling, a boom-boom-ka-boom like distant thunder, like rocks tumbling. The earth shakes and then the sound comes closer, louder; your body is being assaulted. Drums are coming toward you, a purposeful, synchronized stampede, like fifty rhinoceroses, coordinated, all of one mind, as though they have devised a terrible, directed plan for attack and total destruction. It has to be real, you think, but it is a sound you've never heard before. What starts as a quiver of apprehension turns into collective fits of shaking as all of your family and friends wake and tremble with helplessness, all the gumption draining out of your bodies before you even know what is going on. The terrifying synchrony of it, the bone-shaking intensity of it, the sheer loudness. Do you run or prepare to fight? You sit frozen, in awe, paralyzed. What in the world is happening? As they crest the hill, you see them, and for a brief moment before the deafening sounds knock you senseless you see a band of warriors banging on drums in an eerie demonstration of coordinated, malevolent power.

Throughout history, tribes often attacked their enemies stealthily, in the dead of night while their opponents slept. Clever tribespeople, lucky recipients of a bit more cognitive capacity than their neighbors (thanks to random mutation), at some point recognized the power of drum music to incapacitate the enemy, to sap their resolve and simultaneously impassion their own warriors. Each drum tuned slightly differently, skins stretched over wooden stumps, sticks, and rocks knocked together; shells and beads banged, hit, struck, scraped, and shaken: the sound of a well-coordinated, well-practiced single mind. If these invaders could synchronize so tightly to something as nonvital as drumming, that same synchronization put into the service of killing would be so relentless and merciless as to crush even the most formidable resistance.

When Joshua fit de battle of Jericho, it was not melody that made the walls come tumbling down, according to one rabbinic midrash, it was the rhythms of the Hebrew army drum corps. And it was the terrified Jerichoans who themselves opened the walls to the invaders, realizing the futility of putting up a fight, hoping that their conciliatory gesture would eke out a trace of compassion. (It didn't.) At the foot of Balin's tomb in *The Lord of the Rings,* surrounded by dozens of skeletons, Gandalf reads the last entry from the watchman's logbook: "The ground shakes.

Drums . . . drums in the deep. We cannot get out. A shadow lurks in the dark. We can not get out . . . they are coming."

II. It is 7:45 A.M. on a November morning at a high school in Kansas City, Missouri, fifteen minutes before the first period bell rings. Out in back of the school, near the Dumpsters and an abandoned basketball court, a group of students smoke cigarettes. For some it's the first of the day, for others their third. They are not the good students, the star athletes, members of the chess club, glee club, or drama club. They aren't the worst students either, the ones who are being threatened with expulsion or who are being evaluated by a stream of head-scratching school psychologists. These are average students who would otherwise go completely unrecognized and unnoticed by the rest of the school except that they come here several times a day. Most have been in trouble with their teachers or the principal for breaking one rule or another, but nothing serious—being in the hall without a pass, tardiness, late homework—crimes of laxity and neglect, not of violence. The alley where the municipal garbage trucks come has been named Tobacco Road by generations of students at the school. The morning cigarette ritual is followed by the ten o'clock mid-morning recess, lunchtime, and afternoon recess cigarette breaks. They blow smoke rings; they spit. The boys talk about cars they know they'll never own, and Bruce Lee movies they've memorized. The girls talk about older siblings who don't come home at night, with mind-numbing jobs and boyfriends.

None of them has much money, and with the cost of cigarettes approaching fifty cents each, they share a daily concern about where the money for the next pack will come from. But they are generous to anyone who shows up without a cigarette, sharing among one another what they have. When a stranger asks to bum a cigarette, several of the teens offer the outsider the hospitality of a shared nicotine rush. The group are alternately chatty and reflective as the chemicals simultaneously rouse their frontal lobes and calm their limbic systems.

Most of them have beat-up iPods or early MP3 players, but when they're smoking together, the earbuds hang at their sides, and they listen to a boom box or portable player with a speaker built in. "The fidelity is whack," says one, "but least this way we can all hear it together." They tap their feet to 50 Cent—some of them raising the *back* of their heel and pounding it down on the pavement in time with the bass drum. They sing along with all the words to Ludacris, and when Christina Aguilera comes on, the girls do some steps, cop some poses, as the boys try unsuccessfully to feign disinterest. But it's when an old song from thirty-five years ago comes on that one of the girls cranks the volume. Soon the entire group is moving as one to "Smokin' in the Boy's Room" by Brownsville Station:

Smokin' in the boy's room
Smokin' in the boy's room
Teacher don't you fill me up with your rules
'Cuz everybody knows that smokin' ain't allowed in school!

They are singing at the tops of their lung capacity, laughing, and a transformation has come over them. This is *their song.*

Two different scenarios, far removed in time and place. Music, or at least the rhythmic aspects of it, binds the first group together in fear. A seventies song with a distorted electric guitar binds the second group together in defiance. They are two very different types of bonding, but both with an important survival component. Both are bonds of cooperation.

The surprise, predawn attack was a gruesome innovation in prehistoric warfare. The attackers would wait until their opponents were in deep sleep and attack just an hour before dawn, sometimes in complete silence and sometimes with a fanfare of menacing instruments, creating as much noise and mayhem as they could to terrify their victims. By attacking at that hour, they had the element of surprise. By bringing their own torches, they controlled the source of light. By the time the sun came up, they could survey the destruction and collect the spoils.

This is a study of evolution and natural selection in situ. Those bands of early humans who were unable to develop a strategy for fending off such attacks were killed; their genes did not endure in the population. But a few clever humans did develop countertactics—no doubt as a direct consequence of the increased size of their prefrontal cortex, conferred as an advantage by random mutation. These countertactics may well have involved staying awake at night and singing as a way to broadcast, "We're awake, and we're here."

Consider the Mekranoti people of the Brazilian Amazon. They are a small group of hunter-gatherers, indigenous to southern Pará, who have had relatively little contact with modern humans and thus are living their lives in ways that, anthropologists believe, have probably changed very little over the last several thousand years. One of the most remarkable things about the Mekranoti is the amount of time they spend singing—women sing for one or two hours every day and men sing for two hours or more each night. Given their subsistence lifestyle, this represents an enormous investment of time that might be more productively spent gathering food or sleeping. As David Huron writes:

> The men sing every night starting typically around 4:30 in the morning. When singing, the Mekranoti men . . . swing their arms vigorously. The men endeavor to sing in their deepest bass voices, and heavily accent the first beats of a pervasive quadruple meter with glottal stops that make their stomachs convulse in rhythm. Anthropologist Dennis Werner (1984) describes their singing as a "masculine roar." When gathering in the middle of the night, the men are obviously sleepy, and some men will linger in their lean-tos well after the singing has started. These malingerers are often taunted with shouted insults.
>
> Werner reports that "Hounding the men still in their lean-tos [is] one of the favorite diversions of the singers. 'Get out of bed! The Kreen Akrore Indians have already attacked and you're still sleeping,' they [shout] as loudly as they [can]. . . . Sometimes the harassment [is] personal as the singers [yell] out insults at specific men who rarely [show] up." . . .
>
> Like most native societies, the greatest danger facing the Mekranoti Indians is the possibility of being attacked by another human group. The best strategic time to attack is in the very early morning while people are asleep. Recall the insult shouted at men who continued to sleep in their lean-tos: "Get out of bed! The Kreen Akrore Indians have already attacked and you're still sleeping."
>
> The implication is obvious. It appears that the nightly singing by the men constitutes a defensive vigil. The singing maintains arousal levels and keeps the men awake.

The Mekranoti are just one of many examples of people singing to ward off predators or attacking neighbors. It can be seen as the opposite side, a complementary behavior, of the aggressor's use of music. Native Americans often sang and danced in preparation for launching an attack, as did the prehistoric aggressors in the fictional scenario #I above. The emotional and neurochemical excitement that resulted from this preparatory singing gave them the mettle and stamina to carry out their attacks. What may have begun as an unconscious, uncontrolled act—rushing their victims with singing and drumbeating in a vocal-motor frenzy—could have become a strategy as the victors saw firsthand the effect their actions had on those they were attacking. Although war dances risk warning an enemy of an impending attack, as Huron notes, the arousal and synchronizing benefits for the attackers may compensate for the loss of surprise. Combined with the sheer intimidation of witnessing such a spectacle, humans who sang, danced, and marched may have enjoyed a strong advantage on the battlefield. Nineteenth- and twentieth-century Germans feared no one more than the Scots—the bagpipes and drums were disturbing in their sheer loudness, and add to that the visual spectacle, the fearlessness of rank after rank of men wearing skirts. The Romans also feared the Scots in part because of their music—this fear culminated in Hadrian's Wall. Like the Maori in New Zealand with their tattooed faces, open mouths, and outstretched tongues, music becomes a way of shouting at an intimidating a foe, a tactic already well known during Old Testament times, the beginning of recorded history: "Raise the war cry, you nations, and be shattered!" (Isaiah 8:9).

In our own time we have seen the power of such intimidation. Film footage of the marching Nazi army is terrifying to most people (even without knowing that they are Nazis). The synchronous, precise movements of the army suggest a level of discipline and instruction that is beyond the ken of our normal experience. Subconsciously, we realize that if they have mastered such precision in an activity as apparently useless as marching, how much more skilled they might be at the business of killing—hundreds or thousands of soldiers united in synchronous movements, choreographed for death and destruction. That is intimidation, and one of the reasons why patriotic parades often feature infantry marching down the main street of a town. Similarly, the sound of the Mekranoti collectively joined in a loud, middle-of-the-night song, signifies more than the fact that they are awake and vigilant—it indicates strong emotional bonds, coordinated effort among the singers-cum-fighters.

The physiology of singing, as opposed to simply speaking, allows the group to maintain loud voices for a longer period of time because singing uses different throat and diaphragm muscles than speaking. Through singing, especially in harmony, the Mekranoti can give the impression that their numbers are even greater than they actually are. The vocal synchrony further conveys that they are not simply acting as independent entities; its demands also indicate that they are aware and sensitive to the physical and mental states of each member of the group—an awareness that could create a formidable military defense if called to fight.

The primates that we *Homo sapiens* are descended from are manifestly social species. But there are unpleasant by-products of being intensely social and interested in the comings and goings of others: strong rivalries, jealousies, challenges to dominance hierarchies, competition for food, and sexual selection competition for those mates that are perceived to be the most desirable (remember high school?). These social tensions are the primary reason that nonhuman primates are rarely known to travel in groups larger than a few dozen—the social order simply cannot be maintained in larger assemblies.

But larger living groups, if they can be formed and maintained, confer several significant advantages. First, larger groups are likely to be more successful at repelling outside invaders. In a hunter-gatherer society, in which foodstuffs are often difficult to find and secure, the risks of any individual coming home empty-handed are diluted through the actions of many dozens or hundreds of hunter-gatherers; with cooperation, a given individual may come home empty-handed today, but full-armed tomorrow—in either case, the supplies are shared.

The genetic diversity of larger living groups (and the great range in choice of mates) provides a clear evolutionary advantage in that the population will be more resistant to disease and generally more flexible in responding to environmental change. These advantages apply as much to insects and bacteria as to people—but insects so far as I know don't make music, so I'll set them aside. (Bee and ant cities do display a complex social order and flexible roles for the members, but this is not the result of any form of consciousness like our own. Insects do display synchronous, rhythmic behavior which may be musiclike, but it is not really music.)

Humans certainly have overcome the sociobiological limitation on group size found in other primates, establishing living groups in the hundreds (at first, as the size of current hunter-gatherer societies attests), then tens of thousands, and now millions. The United States has nine cities with populations above one million, and China has fifty cities with populations above two million. Imperial Rome had a population of one million around 100 C.E. and ancient Athens a population of about half a million. The Old Testament (Exodus 12:37) refers to six hundred thousand men leaving Egypt during the Exodus (according to Josephus and other historians, dated at around 1500 B.C.E. plus or minus 150 years), and rabbinic teaching estimates the total number of the group that fled across the desert to be above one million. Human living groups in the hundreds of thousands have therefore been around for at least 3,500 years, and groups in the single-digit thousands must be much older.

How did we humans manage to relieve the social tensions that were necessary for the creation first of larger living groups—numbering in the hundreds—and ultimately of society and civilization?

I believe that *synchronous, coordinated song and movement* were what created the strongest bonds between early humans, or protohumans, and these allowed for the formation of larger living groups, and eventually of society as we know it. Throughout our evolutionary history, music and dance typically co-occurred. Rhythm in music provides the input to the human perceptual system that allows for the prediction and synchronization of different individuals' behaviors. Sound has advantages over vision—it transmits in the dark, travels around corners, can reach people who are visually obscured by trees or caves. Music, as a highly structured form of sound communication, enabled the synchronization of movement even when group members couldn't see each other. It allowed for distinctive vocal messages that could be transmitted across territories; for that matter, distinctive whistles and calls could have functioned much as the "secret clubhouse knock" allows identification of people we can't see. Once hit upon, these behaviors would quickly spread, since groups that didn't employ them would be at a competitive disadvantage. As Vernon Reid of the rock group Living Colour said, "In Africa, music is not an art form as much as it is a means of communication." Singing together releases oxytocin, a neurochemical now known to be involved in establishing bonds of trust between people.

In laboratory studies (both in my lab and the laboratory of Ian Cross at Cambridge) two individuals who are asked to synchronize their finger tapping on a desk synchronize more closely than when asked to synchronize with a metronome. This may seem counterintuitive, because the metronome is far steadier in its beat and therefore more predictable. But the studies show that humans accommodate one another's performance, a situation of co-adaptation. They *interact* with one another, but not with the metronome, leading to a greater drive to coordinate. The evolutionary root of this behavior may well be in the coordination of movement, in general, because that serves to facilitate social interactions. If we're walking together and communicating partly through vocalizations, partly through gesture, the interaction is greatly improved if our steps are aligned, if we've synchronized our gait—without it, one person's head is always bobbing in and out of the other's visual frame.

The wartime and hunting aspects are only part of the story. Synchronized movement also made collective tasks much easier to undertake, from hauling heavy objects to building structures to sowing seeds with human-driven plows. And when early humans were engaged in such heavy, manual tasks, looking at others in order to attain motor synchrony would not have always been an option. An aural signal for the synchronization—a repetitive, auditory call with accent structure indicating when certain key movements were to occur—would have made it possible to accomplish a great many physical tasks *as a collective* that would have been impossible individually. Historian William McNeill (author of *The Rise of the West*) highlights the importance of synchronized movement in manual labor:

> Without rhythmical coordination of the muscular effort required to haul and pry heavy stones into place, the pyramids of Egypt and many other famous monuments could not have been built.

Rowing crews on ships, in tight quarters, *had* to synchronize their movements to avoid injury. The same is true on the battlefield, as Sun Tzu wrote in *The Art of War* (fifth century B.C.E.); weapons used at close range could easily harm one's own countrymen if movements were not properly coordinated. And I believe that the muscular coordination was facilitated, prompted, and motivated by song. Songs that were essentially ones of friendship, of social bonding. Where would civilization be without them? William McNeill continues:

> Crops in Sumer depended on irrigation; large-scale irrigation required the construction and maintenance of canals and regulation of how the water was distributed to the fields. . . . To make such feats possible, the scale of human society had to expand far beyond older limits. Villages with no more than a few hundred inhabitants no longer sufficed. And the rich harvests that could be garnered from suitably irrigated alluvial flood plains made it possible to feed the necessary numbers and even to reserve additional labor and materials for the construction of monumental temples in a dozen or more interconnected cities. Cooperation and coordination of effort according to plan were needed to achieve these goals.

Although McNeill's research focuses on the motoric aspects of synchronized movement, he too feels that music was the guiding and binding force behind organizing this cooperation. Work songs ("Whistle While You Work," "Let's Work Together," "Hard Work") do help to pass the time, and may well be a comfort to those singing them, but this is not their fundamental use—primarily, they exist to coordinate movement and cooperative undertakings, to imbue participants with a sense of shared purpose. Track lining songs are special cases of music that unified manual labor by their heavy rhythmic component (one-two-three-heave!). They combined the ancient uses of song with more modern, entertaining features, such as lyrics that often insulted things like the eyesight of the track liner or even the parentage of the crew foreman. Chain gang songs may also fall into this category (when the work being done required synchronous movement) or into the category of comfort, as they helped chained workers to pass the time and increase feelings of kinship with their fellow prisoners.

Synchronized singing and dancing did more than just facilitate the building of large-scale civic structures. They helped build political structures as well. Frictions within a group could be smoothed out by promoting feelings of togetherness. Without explicitly requiring the prelinguistic version of an apology, the strong emotional bonds created by synchronized music-dance allowed both parties to save face and to set their differences aside.

Evolution may have selected those individuals who could settle disputes in nonviolent ways such as music-dance. At a neural level, we now know that the hypothalamus, amygdala, motor cortex, and cerebellum are linked both to movement and to emotion. The basis for this linking goes to the heart of why our ancestors needed to move in the first place: to find food, to escape dangers, and to find mates. All three of these activities are necessary for life, and evolution created links between movement and motivation centers, as opposed to color vision or spatial cognition neural circuits, which are not as closely linked to motivation.

What we call emotions are nothing more than complex neurochemical states in the brain that motivate us to act. Emotion and motivation are thus intrinsically linked to each other, and to our motor centers. But the system can work in the other direction, because most neural pathways are bi-directional. In addition to emotions causing us to move, movement can make us feel emotional. To a neutral observer, synchronized dance appears to be the result of a close relationship between the participants. To the participants themselves, although it may not begin this way, it typically ends up engendering strong feelings of sympathy, caring, and affection. Petr Janata, a neuroscientist and musician, described the strength of these bonds this way: "There are times when I would rather make music and dance with my wife than make love with her—the former can be a more intimate or at least a different type of intimate connection."

Those who march, either in military units or college marching bands, report exhilaration from the activity. Although to an outsider marching drills may seem repetitive and boring, the participants often experience a kind of Zen state of focused attention, readiness, and excitement combined with an almost paradoxical sense of calm—a state called *flow* by the psychologist Mihaly Csikszentmihalyi. A principle of evolution is that in general, if something feels good, evolution must have made it so—evolution must have provided a reward mechanism for synchronized movement and music-making, in the same way that evolution provided mechanisms of reward when we eat and have sex.

William McNeill recalls his days in the infantry:

> What I remember now, years afterwards, is that I rather liked strutting around, and so, I feel sure, did most of my fellows. Marching aimlessly about on the drill field, swaggering in conformity with prescribed military postures, conscious only of keeping in step so as to make the next move correctly and in time somehow felt good. Words are inadequate to describe the emotion aroused by the prolonged movement in unison that drilling involved. A sense of pervasive well-being is what I recall; more specifically, a strange sense of personal enlargement; a sort of swelling out, becoming bigger than life, thanks to participation in collective ritual.

In his insightful history of synchronized military drill, McNeill cites Maurice of Orange, Sun Tzu, Thucydides, and other sources about the effectiveness of marching and the sweeping changes it brought to the battlefield. Some evolutionary theorists might argue that these accounts are too recent (evolutionarily speaking) to be relevant to natural selection, for the good feelings that accompanied such exercises to have been shaped by natural selection. But where threats to life are concerned, natural selection can work its magic in just a few generations. Suppose there are some people who, by virtue of random mutation, enjoy eating dirt. An epidemic of a fatal virus sweeps the world, attacking hundreds of millions of people. It turns out that a particular compound, found only in dirt, kills the virus. Those people who eat dirt would survive and nearly everyone else could be wiped out within only one or two generations.

What we call *instinct* in humans and animals is often nothing more than the product of natural selection at work. Consider house cats. Cats kick dirt or sand or whatever is nearby over their excrement. But it is unlikely that they understand the germ theory of disease and are covering their excrement to minimize contagion. Instead, some ancestral cats had a genetic mutation that triggered the release of certain reinforcing neurochemicals (let's call them "happy juice") when they kicked after excreting. The cats with this mutation were less likely to get sick or to spread disease to their offspring, facilitating this mutation's rapid spread through the genome.

By extension, humans who enjoyed singing, dancing, and marching together so much that they were drawn to it, attracted to it, and practiced it for thousands of hours were those who were the victors in any battles in which such drill conferred an advantage. The strong emotional, even neurochemical pleasure that resulted from synchronized movement may well have had a prehistoric antecedent. Our hunter-gatherer ancestors may have danced around the campfire before and after the hunt. By rehearsing their movements, they gained precision in their actions and were thus more likely to succeed. And taking down a large swift mammal with handheld tools likely required the coordinated movement of many accomplices. Modern army drill is probably an extension of this prehistoric behavior. Music traditionally has been characterized not only by sound but by *action,* and by *interaction* among makers of music-dance.

Humans around the world report not just strong emotional bonding from synchronized, coordinated movement together, but feelings of a spiritual nature—a sense of there being a collective consciousness, the presence of a superior being, or an unseen world that is larger than what we immediately experience. The cognitive psychologist Jamshed Bharucha suggests an explanation for these feelings. The sense of group agency or collective consciousness that one feels when synchronized with others is more than an exhilarating feeling, he says. We feel this exhilaration, which comes from the neurochemical activity described above, and that leads the brain to seek a cause. Attribution—particularly causal attribution—is an automatic and compelling tendency of the brain. In fact, we can't not attribute causes. As we sense a change in our emotional state, we look around to see what's going on in the world that could explain our mood. In the case of group synchrony, we look around us and see all these other people dancing and singing with joy and excitement. In this way, the strange feeling (from the neurochemicals) becomes attributed to something beyond oneself. That's why, in addition to the other advantages of group cohesion mentioned, religions make use of synchronization: It actually enhances the belief in a cause beyond oneself. So it's more than just a good feeling; auditory and motor synchronization can lead to beliefs in forces that transcend the individual, such as societies.

Music and coordinated movement were thus a way of creating meaningful social bonds for these four activities just reviewed: waging war, defending against attack, hunting prey, and forming work crews. A fifth and crucial use of music was for easing tensions within the larger social groups that were forming—group cohesion. Here, music can be traced back even before the emergence of our own species, *Homo sapiens sapiens,* to tens of thousands of years before with a common ancestor, *Homo erectus.* Around the time that *Homo* became bipedal and erect, they left the relatively safe cover of tree living to live on the savannah; the principal advantage was a greatly increased supply of food as *Homo* became hunters, but there were disadvantages as well to be weighed. As Mithen notes:

> Away from the cover of trees, safety can only be found in numbers. . . . There is, however, a cost: social tensions leading to

conflicts can arise when large numbers have to live continuously in close proximity to one another.

Easing these social tensions was not trivial. Among nonhuman primates, this is generally accomplished by grooming one another (picking nits and cleaning the hair of a friend); in fact, the closeness of a relationship between two primates can often be determined simply by the amount of time one spends grooming the other. But with the increased size of living groups—which was necessary for mutual protection—physical grooming of all one's friends and allies becomes impossible. The Oxford anthropologist Robin Dunbar proposed the vocal grooming hypothesis as the origin of vocal communication—the idea that hominids developed vocal communication (music or language) in order to indicate their cooperation and alliance with larger numbers of group members at once.

All over the world and in disparate cultures, human singing is present in two broad styles or forms: strict synchrony and alternation. In strict synchrony, the singers lock their vocalizations in with one another, such as we do in songs like "Happy Birthday" or most national anthems. This requires the ability to anticipate what is coming next in the song (combining cognitive operations of memory in the hippocampus and prediction in the frontal lobes), and then to create what neuroscientists call a *motor action plan*—a specific set of instructions sent to the motor cortex to enable one to sing, drum, or otherwise move the body in time with what others are doing. Part of the evidence that *prediction* processes are involved when we synchronize our singing, hand clapping, or other musical gestures to those of a group is in the small, microtiming errors people make in trying to synchronize: Far more often than not, they are *early* in matching others' musical behavior. This tells us that they're not waiting to hear the next beat before they try to play it; rather, they're anticipating when it will come and preparing a response before it happens. The coordination of activity in these three brain regions (hippocampus, motor cortex, and predictive centers in frontal lobes) would be dependent on the larger prefrontal cortex (than other hominids) that humans evolved.

Alternation occurs when some members of the group deliberately don't synchronize with others, singing either in a round (as when children sing "Row Row Row Your Boat" and some start at a different time than others) or when singing a "call and response" pattern such as in the children's campfire song "Sippin' Cider Through a Straw." Call and response is often found in American gospel music, and is based on an ancient African tradition. Indeed, in sub-Saharan African cultures in particular, this style is considered emblematic of a democratic participation in the music. Call and response is also found in traditional Indian classical music (where it is called *jugalbandi* or *sawaal-javaab* in North Indian classical music), in Latin American music (where it is called *coropregon)*, and in European classical music *(antiphony)*. Alternation in particular requires perspective taking (the first of the three components of the musical brain), and can be seen as an exercise for or predecessor to other more utilitarian cooperative activities. Those individuals who were better able to predict the behavior of others because they could "read their minds" would have had a competitive advantage within the group.

But understanding why it is music and not something else that causes these strong feelings of social bonding remains partly a mystery. Dunbar (and others who followed, including Dean Falk) made the case for why aural bonding would be more efficient than one-on-one physical bonding through grooming behaviors (or through sexual activity as is done by bonobos to promote bonding). Recall that evolution doesn't invent new features from scratch; it doesn't design from whole cloth. Rather, evolution uses structures already in place. Communicative calls and signals were already ubiquitous among the repertoires of nonhuman primates—certain sounds indicated particular types of dangers, the presence of food, and so on. Making such sounds in synchrony would be a clear indication that the group members were paying attention to each other and had a common interest. Among such group vocalizers, those that happened upon a way to induce feelings of happiness, safety, and security in their group mates would have an advantage—these early politicians could cause others to cooperate more with them because they were the source of good feelings.

In a larger context, individuals with social skills would receive many benefits—they would know how and when to get help from others, whom to fight with, whom to trust, and whom to avoid. This emotional intelligence would have given them power over others. Today, in contemporary society, we regard music as a form of emotional communication—perhaps the best one we know. There is no reason to suspect that music functioned differently—although the music itself may have been very different—thousands of years ago. Early humans may have used music to broadcast their own emotional states to others, as well as for the (political) purposes of calming, energizing, organizing, and inspiring.

An important aspect of group cohesion as induced by music-dance is that with larger and larger human living groups, smaller subgroups may form of individuals who feel that their interests are not aligned with those of the larger, dominant group. They may feel as though they lack the power or resources to break out on their own, but that the larger group is not serving their needs. At the dawn of human culture, such a group may have been the elderly, who felt that the social alliances of the young were displacing their own; or a small group of individuals who did not like the current leader and felt mistreated by him. Music has historically been one of the strongest forces binding together the disenfranchised, the alienated.

The high school smokers mentioned at the beginning of this chapter are just one of many such assemblies. In high schools across America there are cliques of "in" students and of "out" students—students who feel marginalized, taunted, or tormented by the stronger, richer, or more popular kids. A common musical interest can provide solidarity for these smaller splinter groups, just as "Smokin' in the Boy's Room" does for the smokers. Gay students may turn to gay anthems such as Lou Reed's "Walk on the Wild Side." The "we" that music binds together can refer to liberals (Nine Inch Nails' "March of the Pigs"), conservatives (Toby Keith's "Courtesy of the Red, White and Blue"), the young (the Who's "My Generation"), the average guy (Primus's "Poetry and Prose"), or the working man (Springsteen's "Working on the Highway"). The free love and sex philosophy of the late sixties and early seventies was celebrated in songs such as Stephen Stills's "Love the One You're With," and those who rejected such notions might have turned to Johnny Cash's "I Walk the Line," and today might be galvanized by Whitney Houston ("Saving All My Love for You") or Jill Scott ("Celibacy Blues"). Cherish the Ladies are a group whose musical mission is to preserve traditional Irish jig music, reels, and airs and provide solidarity for people of Irish descent, especially those far from home; the fact that they all are women positions them as role models for young female musicians.

My mathematics professor at M.I.T., Gian-Carlo Rota, also taught the graduate course in existentialism there in the 1970s and 1980s, and he used to give out buttons that read "Decadence Is Cozy." The message is intriguing: People who do something together that is antisocial or somewhat off-center enjoy a bond. We hear it in the proto-punk classic "Dirty Water" by the Standells. "I'll be down by the river Charles," they sing, along with "lovers, buggers and thieves." What they are saying is "They're actually good people, these river-dwellers, people like us." Much of heavy metal music speaks to people on the fringes of society, the disaffected. Heavy metal lyrics are often a call of togetherness: we (heavy metal fans) are all misfits, but we are bound together in that. A generation was inspired to take drugs, or at least if they were already taking them to feel good about it, by songs such as "White Rabbit" by the Jefferson Airplane, with its call to "feed your head." (Non-drug-users found solace in Paul Revere & the Raiders' "Kicks" or John Lennon's "Cold Turkey.")

The sociologist Tricia Rose points out the role that black women rappers play in binding together other young black women, to give a voice to a segment of society that often correctly feels that their unique concerns are not being addressed. The rappers, Rose writes, "interpret and articulate the fears, pleasures, and promises of young black women whose voices have been relegated to the margins of public discourse."

Patriotic songs—such as the fictional Kazakhstan national anthem that promises the best potassium supply—are a natural extension of the power that music has to define the *we*. This is *our* country, *our* region, *our* group, *our* common interest, *our* football team, even *our* potassium. Although religious leaders have harnessed the power of music to bolster feelings of group solidarity and unity within their sects, their use of music should not be confused with the use of music for ceremonial openings to games and other public events, which is wholly different. Football fight songs and national anthems are essentially songs of social bonding; religion songs have their own character that may *include* social bonding, but this is not their primary characteristic.

Another effective use of social bonding songs is in the political sphere. As I said above, music was used by some early humans to ease social tensions within the group—political schmoozing—and it was also used to allow subgroups, particularly the disenfranchised, to cohere. Protest songs use social bonding powerfully. Whether it's Bob Marley singing "Get up, stand up: stand up for your rights!" or Phil Ochs singing "I Ain't Marching Anymore," Moses Rabbeinu singing "Let My People Go," or Pete Seeger singing "We Shall Overcome," protest songs have an ability to inspire, motivate, bind, focus, and move people to action.

Countless musicians have sung protest songs, and if rock music has a single recurring theme, it is rebellion. One band, the Plastic People of the Universe (PPU), started with no political agenda but is widely regarded as having spurred a revolution in Czechoslovakia. The band started in 1968, the same year that Prague was invaded by Soviet tanks to shut down the liberalization known as the Prague Spring. The new Communist government suppressed free speech, imprisoning many musicians. The PPU were forbidden by the government on several occasions to play, not because of any inflammatory lyric content, but because of their long hair and emulation of capitalist bands like the Velvet Underground and Frank Zappa. (The band took their name from a Zappa song.) In 1970, the government revoked the PPU's musician licenses, which made it impossible for them to get equipment or gigs; they had to play underground concerts to avoid government detection and arrest.

"We were workers," Ivan Bierhanzl, their bassist, says. "For us it was important just to play and listen to our music, and absolutely not to be some heroes." In 1974, the government raided one of their concerts; fans were chased by police with clubs, and some students were expelled, forever ending their academic careers. In 1976, twenty-seven people were arrested at a PPU concert simply for being there. The saxophonist and the lyricist were both imprisoned. Other band members were beaten. A Czech human rights movement emerged, culminating in the nonviolent "Velvet Revolution" ending Communist control of Czechoslovakia. (Tom Stoppard wrote a play about it, which premiered in 2007.)

The unusual thing about the PPU is that they themselves were apolitical and never considered themselves activists, protestors, or revolutionaries with respect to government policy—all they wanted to do was to play their music. But the Communists' actions created a strong support group of activists around the band.

What has been far more common in our lifetime is that protest songs have directly, through their lyrics, addressed slavery, human rights, desegregation, economic injustice, legal injustice ("Hurricane," Dylan's ballad of Rubin Carter), and other social ills. In the past forty years, a particularly large number of protest songs have been antiwar songs, to such a degree that to many people, the phrase "protest song" is synonymous with antiwar songs. And for those of us who grew up in the fifties, sixties, and seventies, disagreements about war created a fissure that seemed sure to drive the country apart. For some people, the moral certainty of peace seemed innate, and protest music gave these people the courage to hold onto their convictions while others around them derided them.

I already had developed antiwar feelings when I was seven years old. I understood World War II—my grandfather had fought in that, and although the war was terrible, the reason for it was clear. A tyrant was trying to kill all the Jews; we were Jewish, and some countries came to our aid. That war made sense. But in 1965 the Vietnam War did not make sense. By October, the United States had sent nearly two hundred thousand marines to Vietnam. The leaves were starting to change color and we did a crafts project with them during art hour at school. Right after recess the teacher had shown us some news reports—young American boys dead on the battlefield. As soon as I got home, I told my mother that we needed to call the President of the United States on the phone and tell him to stop the war. "We can't call the President," my mother said, "he's probably very busy. You know, like when your father is busy at work and we don't call him there unless it is very, very important."

"But this is important," I insisted. "There is no reason that the killing should go on anymore, it can stop today!"

My mother picked up the receiver and called directory assistance to get the number, and then she called the White House. She spoke firmly but matter-of-factly to the receptionist, like calling the President was something she did every day. "My seven-year-old son wants to talk to the President," my mother said, "about the war." She was transferred several times. We got all the way up to the President's chief of staff, W. Marvin Watson. My mother held the receiver against her shoulder. "He said that the President can't talk to you now, he's in a meeting. But he said that he'll pass on the message if you tell it to him." She handed me the phone. He introduced himself, then asked my name and where I lived, and what I knew about the war.

"That the North Vietnamese and the South Vietnamese are killing each other and we went over to help, and now they're killing us. We heard in school about teenagers that went there with the army, and they came back dead. Please—tell the President that he has to talk to them. He has to tell them to stop killing each other. They'll listen to him."

He sighed and I remember hearing that eerie noise of long-distance connections in those days, the clicking and crackling static on the line. He took a deep breath. "We've tried that," he said, his voice cracking. "They won't listen to us. We don't know what to do."

"But tell them," I said, "that we're all just like brothers and sisters. We have to stop fighting!"

"I'll tell the President," he said. "I'll tell him just what you said."

That night I went to bed and heard my parents fighting after I had fallen asleep.

My father and his younger brother were spared both the Vietnam and the Korean wars. My grandfather had been drafted into the army medical corps at the age of thirty-nine, and was away from his sons during four years of World War II, part of that time in Okinawa, where he had engaged in hand-to-hand combat. As a doctor, he had seen the worst bodily destruction imaginable. When his own sons were old enough for military service, he confided to me when I was seven, he intervened—without their knowledge—to make sure that his physician colleagues on the selective Service Board were alerted to medical conditions that may otherwise have gone undetected, and they were classified as 4F, ineligible for service. My father had wanted to serve his country, and had even tried to enlist a year earlier, but my grandfather hadn't let him. My father never expressed remorse or guilt over not having been able to serve, but his principal hobby as long as

I've known him has been reading books and watching films about World War II.

During the 1960s, everyone over the age of seventeen was assigned a draft number, but most people who were in college got deferments. By the time I was eleven, though, the war had escalated. Nixon had just won the White House and the army was starting to take college students, graduate students, medical students, anyone they could get—men in their thirties were being called up. On the nightly news we saw hundreds of flag-covered caskets being unloaded from big transport planes on an airfield in Texas. Now boys in the neighborhood were coming home dead—the older brothers of people we knew. That same year we had to collect butterflies in science class, kill them, and mount them on cardboard. I couldn't do it and my mother had to write a note asking for an alternate assignment. As Vietnam filled the TV news reports every day, my mother saw how worried I was, and at the dinner table one night she said, "Of course if you're drafted, you can say you don't want to go, as a conscientious objector. Or if they don't accept that reason, you can go to Canada."

My father threw his fork down. "He'll do no such thing! If he's drafted, he'll *fight* in the war. It's his duty as an American citizen—his obligation. No son of mine is going to be a draft dodger!"

I had always thought of my father as my protector, that if anything serious ever happened, he would be there to shield me. My mother countered with "He will *not* fight in that war." My parents argued about this all night, long after my little sister and I were sent to bed. Unlike other nights, when we usually fought and called each other names from bedroom to bedroom, this night we spoke softly so that they wouldn't hear us.

"What did Daddy mean? Why was he so upset?" she asked.

"You've seen the war on television," I whispered.

"Yes, between North and South Vietnam," she said. "A civil war." She was now seven herself.

"Daddy said that I might have to go there."

"Nooo!" she said. "You could get killed! He wouldn't say that!"

During the war in Vietnam it seemed as though everybody who was in a position of power or authority in the United States was in favor of it, and those who were most against it were powerless to stop it. This was different from the Gulf War and the Iraq War, in which there was vocal opposition in Washington and very public disagreement from the beginning. To a child, and an antiwar one at that, it gave the Vietnam resistance a kind of David-versus-Goliath feel. There were so many of us against the war, millions by some estimates, but we weren't rich, we weren't in positions of control. The odds seemed overwhelmingly against us. Two of the most important antiwar spokespersons had been assassinated that year, Martin Luther King and Bobby Kennedy. I had seen the Kennedy assassination on live television. My grandfather also died that same year. "We" had tried to take control of the Democratic Convention in 1968, I knew, but we had been held back. Those men, outcasts, rebels at the perimeters of society had tried to get the antiwar agenda heard.

Music was there, songs, to bind together the resistance. I first learned "Where Have All the Flowers Gone?" and "Blowin' in the Wind" as a seven-year-old at a summer camp in the California mountains. A twenty-two-year-old camp counselor brought his guitar to campfire and taught all ninety of us these two protest songs, and we sang them every night for three weeks. As the war escalated, more songs appeared on the radio: "War (What Is It Good For?)," "I Ain't Marching Anymore," "Universal Soldier,"

"Eve of Destruction," and "Bring Them Home (If You Love Your Uncle Sam)." Then there was "Give Peace a Chance," written and performed by John Lennon without the other Beatles. It didn't sound like a Beatles song, but there was that familiar voice, the familiar acoustic guitar rhythms, making the call for an end to the war. Lennon's song was far from the first or even the most popular protest song, but it exploded with musical power and with the raw simplicity of its message. My friends and I memorized even the somewhat tricky lyrics of the verses and sang them in the backseats of station wagons as our parents drove us to Little League practice, Scouts, and to Sunday school. Lennon was on board—he'd step to the head of the line and help lead the antiwar effort. With his charisma and intelligence, maybe now people would listen. This might be the song to do it!

We saw college kids protesting, singing, everywhere. UC Berkeley was just over the hill from where we lived, and the free speech movement, the protests, women's lib, and improved race relations were all bound up into one big cause, into us against them. The songs seemed to hold wisdom, encouragement, and motivation. They were something to play back in your head to remind you that the movement was more than just a thought in your own head, or in the heads of a small group of people you could see. Just *knowing* that there were other people like you throughout the country, hundreds of thousands or millions of protesters, singing the same songs, chanting the same slogans, all with the same goal: The songs provided a strong sense of solidarity.

Then came Kent State, the shooting of four student protesters. This was all we were talking about in my junior high school, going over and over the story in disbelief: The National Guard, the agency formed to protect American citizens in the event of a national emergency, had shot and killed four antiwar activists *just like us.* We had just held our own walk-out the week before, congregating on the football field of our California school, refusing to attend class. For an hour we stood in silence, as did hundreds of thousands of other students throughout the country at the appointed time and place. *What if the National Guard shot us too?*

I was infatuated with the Chicago Seven, whom I considered role models, especially after Graham Nash wrote a song about them, "Chicago."

We all knew the music of Crosby, Stills & Nash. Stills (with his band Buffalo Springfield, which also included Neil Young) had sung his antiwar song, "For What It's Worth," a few years before:

There's battle lines being drawn
Nobody's right if everybody's wrong
Young people speaking their minds
Getting so much resistance from behind
We gotta stop hey watch that sound
Everybody look what's going down

Writing a review of a documentary about the sixties (broadcast in 2007), *New York Times* critic Neil Genzlinger said, "That astonishing song came to encapsulate '60s turmoil so perfectly that resorting to it is a subconscious admission by a documentarian: 'I have nothing to say that Stephen Stills didn't say better in 2 minutes 41 seconds.' Its instantly recognizable two-note opening rings like an alarm bell."

Right after the Kent State murders in 1970, CS&N were in the recording studio with Neil Young. "Teach Your Children" was climbing up the charts and headed for number one. Neil had just written "Ohio" in reaction to the shooting of four student protesters.

"Graham suggested that we release the song right away," Neil Young recalled. "It was his call, because it was his song that was climbing the charts, and we knew that we might not be able to have two songs on the charts at the same time. But he felt it was important to get the song out, and so he sacrificed 'Teach Your Children' for 'Ohio.' That was really something." Nash added, "I had left my group The Hollies over disagreements over which songs to release—I wasn't going to do to Neil what they had done to me." "Ohio" became one of the most moving antiwar anthems; David Crosby can be heard crying at the end of the recording. Many people who grew up in the fifties, sixties, and seventies regarded the leaders of the antiwar movement—whether political leaders or musical leaders—as heroes, taking a courageous stand with the minority, speaking their conscience.

My friends and I spent hours reading everything we could about the assassinations; about James Earl Ray and Sirhan Sirhan; about Kent State. I came to realize that the disagreements about the war were splitting my own house—my own parents, who seemed synchronized on every other aspect of life. Over all this, and the death of my grandfather who had explained everything to me and kindled my young interest in science, I was devastated. But at eleven I could not find a tear for Grandpa Joe, for Dr. King or Senator Kennedy, for the sixty thousand U.S. boys killed, or the three hundred thousand wounded, or for those young college students in Ohio, Allison Krause, Jeffrey Miller, Sandra Scheuer, and William Schroeder. I wanted to cry for them, for all of us. But I was not yet ready.

It eventually became politically untenable to fight the war. It was clear that the United States could not meet any of its objectives.

How much of this was due to the music, the antiwar soundtrack to the protests? It is difficult to say, but music was present at almost every march and rally, in the background of nearly every organizational meeting. At the minimum, it's clear that people at the time at least thought music was helping. But *how* can songs create such changes?

"The arts have power owing to their form and structures," Pete Seeger says. "As I said earlier, good music can leap over language boundaries, over barriers of religion and politics and hit someone's heartstrings somehow. That opens up their hearts to ideas that they might not have entertained if brought in through regular speech."

"I believe in songs, of course," Sting confided to me, "but it's very difficult to imagine that a song would change anything overnight. What you *can* do is to plant a seed in someone's brain, as seeds were planted in *mine* to make me the political animal I am. I think you can sing an idea to a young mind and that young mind may become a political person or a person in power one day and that seed will have borne fruit. Seeger has planted a few seeds that may have borne fruit forty or fifty years later in a subsequent generation."

After visiting Guatemalan refugee camps in the early 1980s, Bruce Cockburn wrote an antiwar song, "If I Had a Rocket Launcher." "Aside from airing my own experience," Cockburn explains, "which is where the songs always start, if we're ever going to find a solution for this ongoing passion for wasting each other, we have to start with the rage that knows no impediments, an uncivilized rage that says it's okay to go out and shoot someone. . . . The idea was to reach a different audience than the politicians by having us go and observe, using the relative visibility that we have to educate the Canadian public to what we had seen and to raise money for projects that OXFAM has in the region."

Here comes the helicopter—second time today
Everybody scatters and hopes it goes away
How many kids they've murdered only God can say
If I had a rocket launcher . . . I'd make somebody pay

I don't believe in guarded borders and I don't believe in hate
I don't believe in generals or their stinking torture states
And when I talk with the survivors of things too sickening to relate
If I had a rocket launcher . . . I would retaliate

On the Rio Lacantun, one hundred thousand wait
To fall down from starvation or some less humane fate
Cry for Guatemala, with a corpse in every gate
If I had a rocket launcher . . . I would not hesitate

I want to raise every voice—at least I've got to try
Every time I think about it water rises to my eyes.
Situation desperate, echoes of the victims cry
If I had a rocket launcher . . . Some son of a bitch would die

Willie Nelson, writer of 2,500 songs including the classic "Crazy" (made famous by Patsy Cline) wrote "Whatever Happened to Peace on Earth?" for Christmas 2003 to protest the Iraq War. "I hope that there is some controversy," he said. "If you write something like this and nobody says anything, then you probably haven't struck a nerve."

There's so many things going on in the world, babies dying, mothers crying
How much oil is one human life worth
And whatever happened to peace on earth

The protest music of the sixties and seventies was often accompanied by marijuana, cocaine, LSD, mescaline, peyote, opium, heroin, plus various amphetamines and barbiturates. To my parents' generation, all of these were "drugs," and they made no distinction between their wildly different effects. Although there were drug addicts on the fringe of society then, as there are now, and people who used drugs primarily to escape problems or responsibilities, or simply to feel good, there were also many people using drugs as a means of self-exploration, gaining insight into their thought processes, or awakening spiritual feelings during a time when organized religion was rapidly waning. Stuck with real spiritual needs and a desire to make sense of the political and social chaos around them, and sensing that the traditional religious institutions had nothing relevant to teach them, they turned to yoga, Buddhism, Ayn Rand, Dylan, Baez, Lennon and McCartney, the Jefferson Airplane, and sometimes to drugs. I never knew anyone who turned to amphetamines or heroin for enlightenment; rather, these were just available as part of the culture. Many figureheads, including Aldous Huxley, Timothy Leary, Ken Kesey, Ram Dass, and John Lennon had used drugs and told of their ability to clarify things, to expand thought, to reveal mysteries about the world and about one's own mind.

The combination of music and drugs proved to be potent, and scientific research has yet to explain it. Each drug acts differently on the brain, and so each has its own particular effects on the musical experience. Some, like cocaine and speed, don't substantially alter consciousness, or the way that music sounds. The hallucinogens, however, change neural firing patterns in ways that can facilitate associations and memories, and fuel imagination. With LSD or peyote, for example, hallucinations may alternate with actual perception, the latter enhanced by connections to new ideas that can be imaginative, insightful, and poetic. Many people have concluded a drug-induced experience by feeling they gained a better understanding of themselves, of their modes of relating to the world and to others; many have said that they felt a strengthened bond with nature. Paul Kantner told me that when the Jefferson Airplane told everyone to take LSD and contemplate nature, "we imagined people like us sitting in a beautiful park (such as San Francisco's Golden Gate Park), surrounded by like-minded free spirits and an atmosphere of love and goodwill. We didn't stop to think that people would be dropping acid in the projects in the inner city, surrounded by filth, crime, and poverty. The drugs had a very different effect on people in those environments."

Clearly whatever effects different drugs had on the brain, there were interactions with the environment, and with differences in each individual's neurochemistry. Brains vary widely from one another in their architecture (that is, the physical size and layout of key structures), the pathways that are available, and their baseline levels of the different chemicals that allow neurons to communicate with each other and ultimately to form thoughts, feelings, hopes, desires, and beliefs. As a neuroscientist acquainted with more than one hundred LSD users, I've come to believe that this particular drug is the most dependent on unobservable factors in each individual's mental makeup. Some people can take hundreds of acid trips and suffer no ill consequences; others take only three or four and are never the same again. Many of these so-called acid casualties have settled on the California coast and I've encountered them in cities like Santa Cruz and Santa Barbara, unable to keep their brains functioning properly.

Music combined with marijuana tends to produce feelings of euphoria and connectedness to the music and the musicians. Δ^9-THC, the active ingredient, is known to stimulate the brain's natural pleasure centers, while also disrupting short-term memory. The disruption of short-term memory thrusts listeners into the moment of the music as it unfolds; unable to explicitly keep in mind what has just been played, or to think ahead to what might be played, people stoned on pot tend to hear music from note to note. Subconsciously all of the usual processes of expectation formation are still occurring (as I outlined in my book *This Is Your Brain on Music),* but consciously, the music creates what many people describe as a time-standing-still phenomenon. They live for each note, completely in the moment.

The proper hallucinogenics, such as LSD, psilocybin, peyote, and mescaline, each have unique effects, but what is common is that they may add to this time-stopping quality a sort of merged sensation or synaesthetic experience: Input from the various sensory receptors seems to merge, and sounds can evoke flavors, smells can evoke touch, and so on. For reasons not entirely understood, but related to action on the seratonergic system of the brain, these drugs also create feelings of unity with those people and things around us. (Serotonin is a neurotransmitter involved in the regulation of sleep, dreams, and moods—it is the chemical system on which Prozac acts.) The ultimate expression of these feelings of connectedness occurs when musicians take hallucinogenics together, tripping together, playing together, and experiencing ecstasy together. This shared neurochemical and spiritual experience has been a sacred foundation of Native American ritual (in both the northern and southern hemispheres) for centuries. In our own lifetime, the Grateful Dead tapped into this and formed powerful connections with those members of their audience who also took LSD, leading to the perception of an intensely synchronized experience between artist and audience. LSD users listening to Grateful Dead music describe the experience using metaphors from electricity: "It feels like I'm *plugged in* to them"; "we're on the same wavelength"; "I get an electric charge listening to Jerry solo." Jam bands such as Phish and the Dave Matthews Band extended the tradition through the 1990s and 2000s.

As ubiquitous as drug taking seemed to be in the sixties and seventies, it really represented a counterculture movement, practiced by a minority of people, and by people who were either in the avant garde or at the fringes of the culture, depending on one's perspective, pro- or anti-drug respectively. I was surprised, then, when my friend Oliver Sacks told me about some of his own drug-induced adventures when I last visited him in New York, adventures that took place in the Topanga Canyon region of Los Angeles in the 1960s. As a neurologist, he had been especially curious about the action of drugs on the nervous system, and wanted to see for himself what the phenomenal experience was like. "I once had a musical synaesthetic dream," he started, "involving musical Pringles potato chips. In my dream, I was eating from a tube of Pringles, and as they crunched in my mouth they would play a symphony or a concerto, each Pringle playing a few bars. That was *without* drugs (although it may have been influenced by previous drug experiences of mine). But with drugs, I typically didn't listen to music, I would sit outside and look at landscapes, or get on a motorbike and go for rides. On those occasions I'd listen to music, I would be sensuously enchanted but often miss the structure of the music."

Oliver described a particular day at a friend's home when the friend was out and music was involved. "I had ingested mescaline and probably some cannabis," he began. "While waiting for the effects to take hold, I put on a phonograph in the living room of the apartment. I was enjoying the music enormously when I became aware of the first hint of the effects of the drugs, a slightly bitter taste in my mouth." Oliver speaks with a British accent, and his voice has the lilting musical quality of a great storyteller. "Suddenly the music was coming from everywhere, not just the speakers, and it drowned out all my other thoughts. I felt at one with a four-hundred-year chain of music leading back to Monteverdi. I saw the most wonderful colors, and my thoughts were freed from their normal patterns. I saw colors I had never seen before, and I felt a great sense of peace. The world appeared to me to be older, more organized than I had previously considered it, and although I am an avowed atheist, I had a strong feeling of a benevolent presence—you might call it 'Einstein's God.'"

Not long after telling me this story, Oliver came to Montreal, where I live, to give a talk to a sold-out audience of eight hundred people in the same lecture hall where I give my cognitive psychology class every winter. He spoke about three of the twenty-nine chapters in his insightful book *Musicophilia,* tales of individuals with various brain disorders that affected their musical experience. The next morning I met him and his executive assistant and editor Kate Edgar at their hotel for a large buffet breakfast. Buffets with Oliver are . . . an *experience.* He takes small portions back to his seat, eats them, and then hurriedly darts back to the buffet with the bearing of a hunter, eyes squinted, hunched over, looking for some hidden treasure. He is usually rewarded for these efforts, bringing back on this day morsels of herring, banana nut bread, or granola that had eluded Kate and me.

We were talking about musical hallucinations, when Oliver jumped up. He returned a few minutes later with a star fruit. Oliver takes nothing in life for granted, finding pleasure in myriad little moments of the day. He cut the star fruit perfectly in half with a brain surgeon's precision, and carefully—admiringly—studied the stellate pattern inside. He then ate the fruit, core and all, while I told him about my own musical hallucinations, which usually occur just as I'm falling asleep (the technical term for these is hypnagogic). Oliver, the rebel drug taker, was particularly interested in a *New York Times* op-ed piece I had recently written, explaining the neurogenetic and neuroanatomical connections between music and movement. The article ended with a tongue-in-cheek call for Lincoln Center to rip out the seats so that people could do what we were programmed to do by evolution: dance to the music. Oliver rocked back and forth in his seat as we discussed the article. "Are you hearing music in your head right now?" I asked him. "I'm almost *always* hearing music in my head!" he answered.

Our breakfast was in the restaurant of the hotel they were staying in, the Fairmont Queen Elizabeth. Oliver mentioned, amusedly, that he had been assigned the John Lennon suite, although he had no knowledge of the history behind that room—that John had stayed there in 1969 as part of a very well-publicized war protest. I remember watching the coverage of Lennon's residency at the Queen Elizabeth from my home in California. I'd lived in Montreal for eight years but never known that the room had been preserved, much less named for Lennon. After breakfast, Oliver—knowing me to be a Lennon fan—asked if I wanted to see the room, so we went up to Room 1742. As soon as we got out of the elevator, I froze in my steps. I recognized the hallway in front of 1742, the same hallway I had seen on *The Huntley-Brinkley Report* on the nightly NBC news.

Kate and I—who are about the same age—kept interrupting each other explaining to Oliver why the room was named the "John Lennon/Yoko Ono Suite." That during the last week of May 1969, Lennon and his wife had staged an event to protest the Vietnam War. Lennon recognized that his celebrity caused reporters to follow his every move, and he wanted to use that for a higher purpose than simply gathering more publicity for himself. He and Yoko came up with the idea of a "bed-in," the honeymoon equivalent of a "sit-in." They would stay in bed for a week and talk to reporters about the war, about peace, and try to use that as a platform for their views. Many reporters mocked them. Some were disappointed, expecting to find the couple making love for the cameras. The couple swallowed their formidable impatience, and played host to a continuous stream of journalists, some prepared to cover a serious story, many more not.

Oliver opened the door and invited us in. My vision of the room took on a kind of split-screen quality, present reality mingled with vivid competing visions of the news reports from forty years ago. Huntley-Brinkley. Cronkite. Peter Jennings. On June 1, 1969, Lennon wrote and recorded "Give Peace a Chance" in this room with Timothy Leary, Tommy Smothers, and others singing backgrounds. The room looks just like it did then, except the mattress has been replaced. The walls have pictures of John and Yoko in that very room during the bed-in. One of the photos, just next to the bed, is in full color, showing John's auburn hair and bushy eyebrows, and the small mole between his eyes, what Yoko describes as one of three moles he had, seldom caught in photographs, this one giving him a Buddha-like third eye. He's cradling his Gibson J-160—I've never seen a photograph of John "holding" a guitar, he always is cradling it as one would a child—and he's drawn caricatures of himself and Yoko on the front with a Sharpie. The color photo is flanked by two black-and-white photos of the couple in bed, talking to reporters. Timothy Leary is in the foreground of one, Tommy Smothers in the other.

Oliver stood in a large sitting room off the bedroom, studying the framed manuscript of "Give Peace a Chance" and a gold record award for five million sales. More photographs of John and Yoko graced the walls in the room, along with a pair of Zen paintings and a still-life bowl of fruit. "I don't know much about popular culture after about 1960," Oliver said, his remark reminiscent in accent and tone of Seth McFarlane's Stewie Griffin. But Kate and I grew up in all of this, and back then had felt that we could change the world just by wanting it to be so.

The two of us are in the bedroom, transfixed by the photographs. *I'm standing right next to the bed where John and Yoko launched this protest, where they sang the song.* I hear it playing in my head. It is anthemic, large, earnest, pleading and yearning for peace, for people to lay down their weapons. The song is recursive in that it refers to itself and to the media frenzy surrounding its recording. The verse is a poke at the reporters who were preoccupied with trying to come up with the right label for the event, with trying to characterize it, while patronizingly ignoring the message *behind* the song.

The words of the refrain, their vernacular quality, sum up the message: All we are saying is "give peace a chance." We've tried everything else—bombing, shooting, napalm, hand-to-hand, air strikes, strafing. Why don't we try *not fighting* for a while and see if that works any better? The message is so simple, from a heart that had managed to keep a sense of childlike wonder about the world even at the age of twenty-eight. He would only live another eleven years.

How can they talk about winning a war when so many people die? Who are the winners and what have they won—the right to have killed so many without repercussions?

I look at the photographs and then at the bed, then at the photographs, and at the radiator, trying to match up the elements of the pictures with the room I'm standing in. I walk to the window and look out—this is what John saw when he was here—the city skyline, the cars below, the buildings across the street. This room was where he wrote the song. As I move from the photographs to the room, studying them, I notice Kate start to tear up.

"I remember where I was when I first heard that song," she says.

"It was full of so much hope," I add. "Lennon believed he could change the world with a song—with that song—he believed that much in the power of music." The song continues to play in my head, but not like an ear worm, stuck in an irritating twenty-second loop, but full, rich, vivid. I hear the percussiveness of his guitar (hastily miked, it sounds thin, more like sandpaper and sticks than the beautiful spectral instrument it is), the clapping of the twenty people in the room, the makeshift bass drum of people clomping their feet on the floor, sounding eerily like mortar fire.

My mind becomes flooded with thoughts I haven't held there in years—the death of my grandfather, of Martin Luther King and Bobby Kennedy, the dead war veterans, the Kent State students, Lennon's own violent death. Standing in Room 1742, Lennon's room, I find a tear for all their lives, and all that they stood for.

CHAPTER 3

Joy

or "Sometimes You Feel Like a Nut"

At one time or another, all of us are filled with a sense of giddy, unrestrainable *joy.* It could be the first sunny day after a long winter; the first glimpse of a recovering loved one we had thought near death; a three-year-old finding the teddy bear that had been missing for months underneath the bed. It can come for no reason at all, just waking up in the morning and feeling good. It can be the result of random chemical perturbations in the brain or external changes in fortune. It is a natural occurrence, an almost unconscious drive to celebrate. In Chapter 1, I mentioned poems and songs celebrating life's little moments—the joy of biting into a garden-fresh tomato, of seeing your newborn take her first steps, of learning for the first time that that special someone loves you back. The natural reaction is to sing, jump, dance, shout—all things that are part of standard music-dance in all societies. Formulating these feelings into a coherent structure makes them into *a* song or *a* dance, but even without such form, they *are* music-dance.

Two great contemporary songwriters, Sting and Rodney Crowell, both feel that the very first songs humans sang must have been songs to express joy.

Sting dropped by to visit my laboratory when his band the Police were on tour in the summer of 2007. I told him about *The World in Six Songs* and he said that he'd like to trade ideas about the origins of music. We met up again that fall in Barcelona, where the Police world tour was continuing.

"I think the first song was just abstract fun with sound. You know, opening your mouth and going 'Aaaaa Ooooo Aaaaa Eeeee Aye!' And once you've developed *that* as a sense of play, or a sense of opening the trachea and breathing—putting stuff out in the atmosphere—then songs come from there. But they're essentially *fun;* it's *fun* to make those sounds. I've noticed that when I sing in concerts, just playing with vowels, there's a shamanic element to sound. Magical—creating a mood of a numinous feeling of connection to everything."

"Sound is different than sight," I offered, "because when you *see* things, it feels like they're out there, but when you *hear* them it feels like they're in *here.*" I pointed to my head.

"Yes—sound joins the inner world to the outer world. What I do in the live Police show a lot is I use a couple of vowels with the audience . . ."

"Ee-oh-oh."

"Yes, I use 'ee-oh' a lot. In Italian it means 'me.' I don't know what *that* says about my psychology! But it's something that the audience clearly gets off on. And it's a simple vowel thing—it creates this bond, it creates this *link* between us all. And you can fill the whole stadium with it. I don't know whether it's meaningless or not, but it definitely has some sort of power—and it's not personal power. The thing itself has a power, has a sense of connection. They're probably the most effective songs, really."

"You do that in 'Every Little Thing She Does Is Magic,' and in 'De Do Do Do, De Da Da Da.' "

"Yes, and you'll hear me do it tonight in 'Walkin' on the Moon' and another song too. It's powerful. People relate to it; they feel the joy and the magic. So getting back, I think that the first 'song' was a caveman just playing with sound and other people joined him and they liked it—it felt good. And he would have been moving—there can't be any music without movement."

"I think songs spring from the surrounding environment, and are archetypal," Rodney told me. "I think the first song was probably the caveman equivalent of 'You Are My Sunshine.' If you go back to cavemen—if they were going to illustrate life using sound—the sun would be the *first* thing they'd want to sing about, and they'd sing about if joyfully. . . . You give music to what you feel in your senses, to your sensibilities, your perceptions; I would walk out on a day like today and the first thing I would notice was what the sun was doing. That would be my first song. Of course the Jimmie Davis song 'You Are My Sunshine' is about a person, but for a caveman, a song about the sun is really a song about Creation. The sun is this ball of fire; it's light and heat and ultimately represents survival. What we do as songwriters, as creators, is to acknowledge our surroundings. We try to do with music what painters do with paint."

Joy songs are found in every corner of human experience where we might look for them. My grandmother—my mother's mother—was an immigrant to the United States from Germany. Like many, she came to the U.S. to escape terrible tyranny and oppression, leaving a country where her parents were shot by soldiers right in front of her eyes as they sat in their living room. She told me when I was eight that she woke early every morning to sing "God Bless America." She sang it for me every time I saw her, her voice trembling through her thick accent, her body shaking and overflowing with joy and thankfulness that she had been saved, that she had lived long enough to see her freedom and the freedom of six grandchildren.

When she turned eighty, my mother and I bought her a little eighty-dollar electronic keyboard. She didn't know how to play it, but we attached pieces of masking tape with numbers on them to show her the correct order of the notes for that song. Within six months she had learned to plink it out, and she played and sang "God Bless America" every morning until she died at ninety-six. She sang it as if her life depended on it. And maybe it did. She eventually learned how to play a rudimentary harmony. I wonder if having music prolonged her life; it certainly made what time she had more purposeful and meaningful. Neuroscientists have recently found that playing music can modulate levels of dopamine, the so-called feel-good hormone in the brain. The exact mechanism by which this happens is not well understood, but the secretion of feel-good chemicals in the brain in response to playing and listening to music points to an ancient and evolutionarily advantageous connection between music and mood. As I showed in Chapter 2, those of our ancestors who were able to communicate with music, and who *enjoyed* musical communication, may well have been at a distinct advantage in their ability to forge social bonds, diffuse tense social situations (that might otherwise have led to combat and death), and convey their emotional states to those around them. What we know for certain is that increases in dopamine lead to elevated mood and help to boost the immune system. The joy of playing music, the sound and sense of mastery, may well have played Grandma into her late nineties.

Her husband, my grandpa Max, bought a large conga drum on a trip to Cuba before I was born, and whenever he came over, he would sing songs to Grandma while beating his two hands on the skin of the drum. I know now that they had a difficult and stormy marriage, but all I remember of it is the look that came on her face when he sang to her and played that drum—a look of utter enchantment and forgiveness. He would sing—badly, by everyone else's account—and she would melt, the lines in her brow relaxing until she would start to laugh, and gently stroke his hair. The ebullient joy he brought to his singing and drumming were infectous and disarming.

Joy songs today are found everywhere from the scat singing of Ella Fitzgerald to that of the Azerbaijani singer Aziza Mustafa Zadeh, from "Zip-A-Dee-Doo-Dah" to Ren & Stimpy extolling the virtues of their favorite toy, in "Log Blues":

What rolls down stairs alone or in pairs
Rolls over your neighbor's dog?
What's great for a snack and fits on your back?
It's Log, Log, Log!

It's Log, Log, it's big, it's heavy, it's wood.
It's Log, Log, it's better than bad, it's good!
Everyone wants a Log! You're gonna love it, Log!
Come on and get your Log! Everyone needs a Log!

In fact, advertisers in the last thirty years have become the chief creators and purveyors of pure joy music, as they vie to have our good moods and positive neural chemistry associated with their products. Based on their jingles, we are meant to believe that pure, undiluted joy will come to those who eat Peter Paul candies ("Sometimes you feel like a nut—sometimes you don't"), drink Pepsi ("It's the Pepsi generation, comin' at ya, goin' strong/ Put yourself behind a Pepsi/If you're living, you belong"), or drive a Chevrolet ("See the U.S.A. in your Chevrolet!"). The company that manufactured the children's toy Slinky turned a *spring* into a perennial children's favorite, largely on the basis of their catchy song! (And it was this song that Ren & Stimpy were parodying when they tried to show that even a *log* could be a fun toy if given the right song.)

The prominance, if not dominance, of joy songs in the commercial sphere today points to a plausible role for them during evolutionary time frames. The group member who could make others feel good, either through grooming, sexual activity, providing more food, and so on, was one who became valued and could ascend to the position of group leader, in which case the community would work to meet his needs for him. Communication by sound allowed a potential leader to spread his influence around to many more at a time than could be done by one-on-one grooming.

Confucius reportedly said, "Music produces a kind of pleasure which human nature cannot do without." Two thousand years later, Nietzsche—who on most other matters couldn't be farther removed from the ideas of Confucius—wrote "My melancholy wants to rest in the hiding places and abysses of perfection: that is why I need music." Music and health are intimately related in human history, from shamanic healing to "witch doctors," from the Hebrews to current-day programs of music therapy. King David played the harp to relieve the stress of King Saul (Samuel I, 16:1–23), and the ancient Greeks (in particular Zenocrates, Sarpander, and Arien) used harp music to ease the outbursts of people with mental illnesses. Music-as-therapy was also employed by such geographically disparate cultures as the ancient Egyptians, Indians, and Native Americans. Health benefits have been described whether patients sit and listen to music, improvise tunes, write songs, discuss lyrics, perform compositions, or actively participate in the production of music. Music is claimed to be beneficial for patients of any age, ethnic or religious background, or stage of illness.

Before getting too swept up in this, however, let's look at the science behind it. Scientists are understandably skeptical of claims that are not properly substantiated, or of findings that don't reveal the mechanism underlying the observation. For example, we know that singing releases endorphins (again, a "feel good" hormone) but *why* is not known, and this lack of a causal understanding makes many scientists uncomfortable about the connection between singing and endorphins. Could it be primarily an artifact of the breathing involved? If it is specific to singing, why would this be so?

As the cognitive scientist Gary Marcus reminds us, the brain has been shaped by evolution and adaptations that arose independently of one another to solve specific problems. Among other things, brain adaptations occurred in order to help us reach core goals of finding food, avoiding disease and predators, conserving energy, circumventing danger, seeking physical comfort (including homeostasis to protect our bodies and organs), encouraging reproduction, and ensuring the successful maturation of offspring. To this list David Huron adds adaptations such as the ability to anticipate the future, solve puzzles, distinguish animate from inanimate objects, identify friends and enemies, and avoid being manipulated or deceived.

The way in which the brain encourages us to pursue adaptive goals is that it has assembled a system of rewards and punishments through evolution. These rewards and punishments are affected through our emotions—what I defined in Chapter 2 as neurochemical states in the brain that motivate us to act. In other words, emotions and motivation are two sides of the same evolutionary coin. We experience positive or negative emotions as a consequence of the particular neurochemical soup that is in our brains at any particular time, and these emotions cause us to act (or refrain from acting) in particular ways. Pain is one of nature's ways of preventing us from doing things that are harmful; pleasure is a way to motivate us to undertake actions that will increase our adaptive fitness—reproducing, eating, sleeping, and so on. Recall Daniel Dennett's argument that we don't find babies cute because they are *instrinsically* cute; rather, we are the descendants of people who nurtured and protected their babies and were intrinsically rewarded (through *cuteness* detectors, let's say) for doing so. If we find the smell of rotten food or feces disgusting, it is not because they really and truly smell bad (in any objective sense), but because those of our ancestors who had a genetic mutation that caused them to avoid these things (by co-opting their olfactory sense) were those who fared better in the genetics arms race to pass on their genes. When we find something pleasurable or displeasurable, it is often because tens of thousands of years of brain evolution have *selected* for those emotions; natural selection has favored them because they led to motivational states that served our ancestors well in the competition for resources, mates, and health.

In "Heard It Through the Grapevine," when Marvin Gaye sings:

People say believe half of what you see
Son, and none of what you hear
I can't help but bein' confused—if it's true please tell me dear

he is expressing skepticism about his relationship, an evolutionarily adaptive trait if experienced under reasonable circumstances and in reasonable amounts, because it is *maladaptive* in the long run for a male to care for a female who might be mating with another man—in effect, the song's protagonist might be tricked into sharing his resources with a child that is not his.

The emotional flip side of the song might well be "Suspicious Minds" (as recorded by Elvis Presley):

Why can't you see
What you're doing to me
When you don't believe a word I say?

We can't go on together
With suspicious minds
And we can't build our dreams
On suspicious minds.

Too much suspicion erodes the foundation of trust necessary for most long-term human cooperative ventures. The crucial point of all this is that suspicion, trust, conciliation, and even love—indeed all emotions—are products of evolution by natural selection. David Huron sums it up eloquently: "The only emotions we experience are emotions that have arisen through natural selection as adaptations that enhance survival. Jealousy, embarrassment, hunger, disgust, ecstasy, suspicion, indignation, sympathy, itchiness, love—all are adaptations. . . . Nature doesn't build mental devices whose purpose isn't related to adaptive fitness."

How does music fit into the pleasure and fitness story? There is no debating that music can induce pleasure, and that those same chemicals help to boost the immune system. But the neurophysiological machinery involved in pleasure is highly complex. Although there do exist discrete "pleasure centers" in the brain, dozens of neurotransmitters and brain regions contribute to feelings of pleasure. On the research side, while there are many reported cases of music having a positive, and sometimes extraordinarily powerful, effect on the ill, there have been few true experiments performed to document this. The sheer number of anecdotes is impressive, but does not constitute scientific proof, any more than the sheer number of reports of alien abductions constitutes proof of little green men conducting grisly experiments aboard metal saucers that hover above Kansas and Nebraska. (Why the alien abductions seem to happen with greater frequency in the Plains states is another mystery calling out for an explanation.)

Scientists are in the business of wanting proof for everything, and I find myself caught somewhere in the metaphorical middle on this issue. As a musician, I'm reminded on a daily basis of the utterly ineffable, indescribable powers of music. I've also witnessed the healing power of music firsthand. In old people's homes and convalescent hospitals, when people have lost their memory due to Alzheimer's disease, stroke, or other degenerative brain trauma, music is one of the last things to go. Old people who are otherwise unable to remember the names of their spouse or children, or even what year it is, can be brought arrestingly back to focus by hearing the music of their youth—songs that they sing along with, tap their feet to, and can remember all the notes and lyrics of. I've seen patients who could barely move, people with Parkinson's who couldn't walk, who can suddenly walk, trot, dance, and skip as soon as I start playing Glenn Miller or Artie

Shaw on the rest-home CD player. There are reported cases of children with Down syndrome who can't tie their shoes unless the sequence is set to music.

This ineffable power of music shows up not just in listeners but also in creators of music. The great songwriters and improvisers talk about not so much *creating* music, but having it written *through* them, as though the music comes from outside their bodies and their heads, and they are merely the conduit for it. Many great musicians, particularly in Third World cultures, reach a state of total ecstasy, a trance state, while playing music, in which their minds and bodies seem to be possessed by otherwordly forces. I've also felt this, whether improvising onstage at the Santa Monica Civic Auditorium with Mel Tormé, or writing incidental music for the film *Repo Man*. In describing the writing of one of my favorite songs of hers (1990's "What We Really Want"), Rosanne Cash told me "It felt like I had just stuck up my hand and caught the song like you'd catch a ball in a catcher's mitt—like it was out there all along waiting for me to grab it." Our scientific theories have to be able to reconcile this common experience and the strong intuition that music is—dare I say it?—*magical.*

On the research front, many of the studies on the effectiveness of music therapy were not performed according to rigorous scientific standards, and so their claims remain unproven. This situation parallels the unfortunate history of psychic research. One of the most crucial features of a rigorous experiment is the use of the comparison or control condition. In essence, we need to ask the following question: If there were *no* effect at all from the thing I'm studying, would this outcome have happened anyway? Too many music therapy experiments had inadequate controls, meaning that we aren't shown what might have happened to the people in the experiment *without* music therapy.

Consider, for example, that out of twenty people who complain of headaches, a certain number are going to get better *anyway* if you just wait a few hours. If we play soothing classical music to twenty people with tension headaches and six of them say their headaches went away, we don't know if some or all of those six headaches would have just gone away on their own. A control group in such an experiment should be similar in all respects to the people we're studying, and get all the same treatment except for the one thing we're interested in. If we give ten headache sufferers classical music and have them sit in a comfortable, sunlit room, and we give another ten headache sufferers *no* classical music but have them sit in an uncomfortable, darkened room, we have made the mistake of varying three parameters at once: We can't determine which of those parameters had the effect.

In one published study on music therapy, a group of Korean researchers took stroke survivors and gave them an eight-week program of physical therapy that involved synchronized movements to music. The patients recovered a wider range of motion and flexibility compared to a control group. So far so good. But the control group had *no* therapy—no personal contact, no movement (with or without music), no one rooting for them or telling them that they would get better. We don't know now whether the benefits to the first group came from the music, the movement, or simply the good feeling that came from knowing that a medical professional was looking out for and following their progress. Health improvements have been observed with far less.

I mentioned psychic research, and it is an irresistible subject.

Some of the most interesting experiences I've had in my entire life were serving on review panels for scientists who had applied for funding to undertake research on psychic phenomena. I was asked to review their pilot data, findings from preliminary experiments that they felt showed evidence of psychic phenomena. In every single case, a lack of careful scientific controls rendered the data uninterpretable. In one study I reviewed, the person who was "reading minds" could only answer questions correctly if the experimenter already knew the answer and was allowed to interact with the "mind reader." If the experimenter was silenced, the effect went completely away. I don't think that the pair were trying to hoodwink anyone, but a parsimonious explanation—and a review of the experimental transcripts—suggests strongly that the experimenter was providing subtle, if unconscious clues to the "mind reader."

What I found so interesting was the tenacity with which people, even trained scientists, held onto their beliefs about the supernatural when confronted with evidence that the experiments were flawed. First, here's an example of how probability theory pertains to psychic claims. Suppose you have a standard deck of fifty-two playing cards. A friend of yours tries to guess the suit (hearts, clubs, diamonds, or spades) of each card—you can either look at the card (and try to psychically transmit the information) or you can keep it turned down until after your friend guesses. Now without working through a formal mathematical/probabilistic treatment of the problem, it should be clear that if your friend only guesses and has no psychic ability at all, she will guess a few of the cards right. In fact, in the long run, she will tend to get 25 percent of them right. It is the function of probability and statistics to help specify just how many she would have to get right for us to know, with reasonable certainty, that she wasn't guessing.

While reviewing one such experiment in Silicon Valley, California, a very complicated experiment with many different facets, I pointed out to the lead research scientist (who held a Ph.D. in physics) that the chances of guessing a right answer in his psychic experiment were one our of four and he agreed. I pointed out that his best subject, after testing twenty people, had only gotten one out of four correct. He agreed to that. I suggested that she might have been only guessing.

"No!" he insisted. "She *told* me that she was really concentrating."

I asked what his explanation was that she got a meager 25 percent correct, the same number that would have been guessed by a machine generating random numbers.

"She showed her psychic powers on 25 percent of the trials—what more do you want?" he demanded. He was getting agitated now. He started to speak very slowly. "Psychic powers can come and go like anything else. Even Artur Rubinstein doesn't play Beethoven perfectly every time he sits down at the piano." He knew my weakness.

"She had her powers on those 25 percent of the trials. And on the other 75 percent of the trials, the ones she got wrong, *those* are the ones where she was guessing!" I held my ground. If she had been truly guessing on those 75 percent of the trials, she would have gotten 25 percent of *them* right. He would have none of that. He had now stood up from the table and was red in the face with fists clenched, and his knuckles were turning a kind of ghostly whitish yellow. He seemed to be trying to stare me down.

"I have an idea," I said at last. "Why don't you have your subjects *tell* you which trials they're guessing on, and which they *really, really know.* If your subject can get 25 percent of the suits right and can say ahead of time, before she gets any feedback, that those and only those trials are the ones where she is using psychic power, then I think we might have something."

"We've done hundreds of experiments already using our existing method. We have all the data. Why should we go back to the experiments again just to satisfy one $@%* like you? I *know* that she has psychic powers, *she* knows it. Why can't *you* admit it, Dan? Why do you have to be so *negative* all the time!"

The professional magician and skeptic James Randi has offered a one-million-dollar prize to anyone who can prove the existence of psychic phenomena, anytime and anywhere, anyone who can read minds, predict the future, influence the toss of a coin, or divine what playing card is about to be turned up, without using magic. No one has even come forward to try to claim the prize, but the money is in a certified escrow account, there for the taking. A researcher must simply follow the protocols designed to distinguish flimflam from fact.

Which brings me to the healing power of music. There are mountains of data on the effectiveness of music on illness, but not all reliable or reputable. Trying to separate the good from the bad would be enough work to earn some enterprising young investigator a Ph.D. thesis. If I sound skeptical or negative, I do not mean to denigrate the many fine music therapists who *are* helping people. Indeed, the American Music Therapy Association is just as interested as I am in weeding out those who are fakers, exploiters, and just plain incompetent. By the association's own definition, music therapy is the "*evidence-based* use of music interventions to accomplish individualized goals within a therapeutic relationship by a credentialed professional . . ." [emphasis mine]. Certified music therapy is used for pain and stress reduction, motivation, anger management, as an adjunct to physical therapy in the case of motor difficulties, and for a variety of other purposes.

In just the past three or four years, however, an emerging body of evidence is pointing scientists in new directions. There have only been a dozen or so careful, rigorous studies and so I don't want to overstate the case, but they seem to point to what the ancient shamans already knew: music—and particularly joyful music—affects our health in fundamental ways. Listening to, and even more so singing or playing, music can alter brain chemistry associated with well-being, stress reduction, and immune system fortitude. In one study, people were simply given singing lessons and their blood chemistry was measured immediately afterward. Serum concentrations of oxytocin increased significantly. Oxytocin is the hormone released during orgasm that causes us to feel good. When people have orgasms together and oxytocin is released in both, it causes them to feel strong bonds toward one another. "I feel good/I knew that I would/I got you." You can see how *this* would be an evolutionary adaptation. Because the act of lovemaking (at least in the pre–birth control world) often led to pregnancy, it would be adaptive for the man and woman to feel a sense of connection, because that would increase the chances that the man would help raise the child, in turn significantly increasing the child's chances of survival. Significantly, also, oxytocin has just been found to increase trust between people. Why oxytocin is released when people sing together is probably related evolutionarily to the social bonding function of music we saw in the previous chapter.

Looking beyond mental health to physical health, immunoglobulin A (IgA) is an important antibody that is needed for fighting colds, flus, and other infections of the mucous system. Several recent studies show that IgA levels increased following various forms of music therapy. In another study, levels of melatonin, norepinephrine, and epinephrine increased during a four-week course of music therapy, and then returned to pretherapy levels after the music therapy ended. Melatonin (a naturally occuring hormone in the brain) helps to regulate the body's natural sleep/waking cycle and has been shown effective in treating seasonal affective disorder, a type of depression. It is also putatively linked to the body's immune system because some researchers believe that it increases cytokine production, which in turn signals T-cells to travel to the site of an infection. Both norepinephrine and epinephrine affect alertness and arousal, and activate reward centers in the brain. All this from a song.

Music listening also directly affects serotonin, the well-known neurotransmitter that is very closely associated with the regulation of mood. (Prozac and a number of other recent antidepressants act on the serotonin system and belong to the class of pharmaceuticals called SSRIs, *selective serotonin reuptake inhibitors.*) Seratonin levels were shown to increase in real time during listening to pleasant, but not unpleasant music. And different genres of music caused different neurochemical activity! Techno music increased levels of plasma norepinephrine (NE), growth hormone (GH), adrenocorticotropic hormone (ACTH), and β-endorphin (β-EP) concentrations, all chemicals closely associated with improvements in human immune function. Techno was also shown to increase cortisol levels (not good for the immune system, but outweighed perhaps by the other increases), while meditative music decreased cortisol and noradrenaline. In the same study, rock music was shown to cause decreases in prolactin (at least in this group of techno-loving listeners), a hormone associated with feeling good.

We all suffer from stresses today that are very different from the stressors experienced by our ancestors, those very ancestors whose lifestyles caused the changes in DNA that we call evolution. When changes in lifestyle or environmental conditions created a subset of people who were better adapted to those early conditions, natural selection teaches us that those people were the ones who survived to pass on their DNA. This whole process can take a lot of time, thousands or tens of thousands of years. In other words, many parts of our DNA were selected for by evolution to cope with the world the way it was five thousand or even fifty thousand years ago. As biologist Robert Sapolesky points out, we are living in bodies and thinking with brains that were designed to solve problems that almost none of us has today.

In ancestral time periods, if a lion approached us, we became stressed. Cortisol levels shot up. Our amygdala and basal ganglia set us running—or at least those of us who managed to survive. (Many of those early humans who, for one reason or another, didn't run or otherwise escape the lion didn't live to tell about it or to have children.) Running uses up glucose and helps us to "burn off" the cortisol our adrenal cortex produces. Today, though, when our boss yells at us, when we have a big exam that we haven't prepared for, or when someone cuts us off while driving, our adrenal cortex still produces cortisol—the stress hormone—but we don't have an opportunity to burn it off. Our legs and shoulders tense up to run in accordance with an ancient evolutionary formula, but . . . we sit there. Our shoulder muscles stay tense but we are not swinging our arms, and so there is no release.

All that cortisol temporarily interrupts our digestive system—a body in flight needs to allocate its energy to movement and agility, not digestion—and so today, following stress that doesn't require literal fight or flight, we end up with stomachaches, gastroenteritis, ulcers. Increased cortisol is associated with decreases in production of IgA, and so our immune system takes a hit. (This is why people who are stressed are more likely to get sick.) In contemporary society, increased cortisol levels (and decreased IgA) have been found in experiments conducted during some of the most psychologically stressful situations humans face: students before exams, professional coaches during athletic events, and air traffic controllers during their duty cycle. Getting tense in the face of a threat was adaptive for our ancestors; it is maladaptive for us when those stressors are long-term, chronic, and don't require an acute physical response.

So cortisol suppresses our immune system temporarily, marshaling all the resources it can for the task at hand (or at foot as the case may be). This may well be one of the reasons why we move our feet or snap our fingers when we hear music. To the extent that music activates our *action* system—motor sequences and our sympathetic nervous system—our hands and feet become the instruments of that activation. Through these movements we burn off excess energy that could otherwise be toxic. In a sort of neurochemical dance, music increases our alertness through modulation of norepinephrine and epinephrine and taps into our motor response system through cortisol production, all the while bolstering our immune system through musical modulation of IgA, serotonin, melatonin, dopamine, adrenocorticotropic hormone (ACTH), and β-endorphin (β-EP). Some of the energy we feel during music playing and listening is then expended in the increased mental activity (the visual images that many people report accompanying musical activity, or other mental activity such as planning, ruminating, or simply aesthetic appreciation). Finger snapping, hand clapping, and foot tapping help us burn off the rest, unless of course we actually get up and dance, perhaps the most natural reaction, but one that has been socialized out of many Western adults.

But *why* does music—a collection of sounds—tap into all these chemical and activity centers of the brain? What might have been the evolutionary benefit? First, it is important to reframe the question as concerning music-dance—not as simply a collection of sounds we make or perceive, but as an integrated cross-modal experience of movement, synchrony, sound, and perceptual organization, and again, this is because music and dance were virtually inseparable across evolutionary time scales. Second, the musical brain didn't evolve in isolation from other mental and physical attributes. In other words, early or protohumans didn't suddenly end up with music-dance and no other cognitive strengths. The musical brain brought with it all the facets of human consciousness itself. In addition to social bonding, fundamental to the experience of early humans was communicating their emotional states to others—the expression of joy through music-dance.

Unrestrained joy usually accompanies a positive outlook. In situations where success isn't assured, those with a positive outlook are more likely to achieve it than those with a defeatist attitude. Of course there is a delicate balance. As Barack Obama said during the 2008 presidential campaign (quoting the German Protestant theologian Jürgen Moltmann, whose words have also been used by the Catholic Church in offical writings), "Hope is not blind optimism." An overoptimistic person is going to experience a large number of failures and find he has expended considerable energy for no rewards. On the other hand, the defeatist (or pessimist) is going to forgo activities that in many cases would have yielded a substantial positive payoff. The best adaptive strategy for hunting, foraging, or even mating has been shown to be the adoption of an attitude that is slightly over the halfway mark, on the optimistic (joyful) side of realistic. Music has a twofold role to play here, physical and mental. First, joyful music makes us feel better, it pumps us up, picks us up out of the doldrums. Second, joyful music can serve as a model—we look to the creator of that music as a mental inspiration and try to be like him or her.

The clearest case of the evolutionary advantage of optimism might be the caveman who is uncertain whether that glance he just received from a cavewoman was a "come hither" or a "get lost" look. The caveman who walked away may well have lost an opportunity gained by his rival who treated that ambiguous look as at least worth investigating. As a species, we have evolved a healthy distrust for people who are *too* optimistic—they may be deluded nutjobs—and we've evolved a reasonable attraction to people who are self-confident and optimistic—after all, they may know something we don't, and things might just work out well for them. "I'd do well to hitch my wagon to his," we think. The optimist thinks a brewing conflict might be solved by diplomacy. The pessimist thinks fighting is inevitable, and those thoughts may bring about his own destruction. Our brains evolved the responses to joyful music making that they did because joy can be a reliable indicator of a person's mental and physical health.

In his groundbreaking book *Sweet Anticipation,* David Huron spells out how the musical brain might have helped to prepare humans for survival. To what he has already written, I would add that it also served to relieve stress through the release of the very same neurochemicals that helped to ensure survival in hazardous, ancient environments. The ideas are important enough that I think they're worth repeating here in some detail. Huron's thesis is built around a five-step process that he calls ITPRA. I present here a four-stage version of his model, which I think is more parsimonious.

The core idea is that music gives the brain opportunities to explore, exercise, play with, and train those mental, physical, and social muscles necessary for the maintanance and formation of society as we know it. It offers a safe forum in which we can practice and hone skills that are vital through the life span. In my stripped down version (with apologies to David Huron), TRIP stands for Tension, Reaction, Imagination, and Prediction.

Imagine, David invites us, that we witness a lion attack. The next time we see a lion, we will understandably experience Tension. (If we didn't, we might act complacently and end up as his lunch.) The tension begins a cascade of electrochemical processes in our brain and spinal cord, causing us to React. If that reaction allows us to survive, we may then spend some of our time Imagining—recalling the event in our mind's eye (and ear) and planning appropriate reactions in the event of a future attack. Part of this process might entail imagining what future confrontations might look like, how we might Predict a possible attack under different situations.

Now learning about the world by narrowly escaping from lions, rattlesnakes, or angry neighboring tribespeople is not the most efficient way to acquire survival information. Indeed, the topic of Chapter 5 is how particular kinds of songs—knowledge songs—can encode and embed such essential information in a way that is easily remembered and transmitted across time. But before there can be knowledge songs, there must be music, or at least the cognitive foundations for it, an adaptive motivation for the musical brain to come into existence in the first place. This is where music meets TRIP. What if we humans had a way that we could invoke tension in a safe, nonthreatening context, react to it, imagine new forms of tension and our reactions to those, and prepare a repertoire of responses, all from the safety of the campsite, from the safety of our minds? Music doesn't have to be the *only* adaptation that provides this; it only needs to be a *plausible* adaptation, even one among many possible, for this theory of its origins to hold.

Music theorists since Aristoxenes and Aristotle, through Leonard Meyer, Leonard Bernstein, Eugene Narmour, and Robert Gjerdingen, have talked about tension as being one of the core properties of music. Virtually all theories of music assume that musical tension changes during the course of the piece, involving increases and decreases in a cyclic dance of tension and release. In a paper published a few years ago in the journal *Music Perception,* my students Bradley Vines (now a research scientist at UC Davis) and Regina Nuzzo (now a professor at Gallaudet University in Washington, DC) and I described this property of music in terms of physics—Newtonian mechanics specifically. We compared music to a coiled spring like the one you might have attached to your garage door. Pull or push the spring and it tries to come back to its resting position.

Musicians and composers are speaking metaphorically of course when they talk about tension and release in music. But across many studies, the metaphor seems to have consistency of meaning, even among non-musicians and across very different cultures. We seem hardwired to "get" the relationship, however metaphorical, between musical tension and the tension that we feel in physical objects (like springs), in the body (as in muscles), and in social situations (as at high school dances). These common life experiences among humans cause a convergence of meaning for "tension" and "release" across individuals when referring to music. The cognitive psychologist Roger Shepard reminds us that the human mind co-evolved with the physical world in such a way that it has incorporated certain physical laws. No human infant is surprised when objects fall downward—gravity has become incorporated into the hardwiring of the human brain from birth. Indeed, infants as young as a few weeks show surprise when objects are experimentally manipulated to "fall" up, or when one billiard ball hits another and the second does not move appropriately.

In general, tension tends to built up during music to a peak, after which the tension is released and subsides, often rapidly. This is what gives us that "aahhhhh" feeling at the end of a piece of music. Symphonies (from the standard-practice period of classical music), perhaps more so than other musical forms we enjoy today, are particularly formulated to create this sense of dynamism, of tension and ultimate, rewarding release in the last few moments. In performances of Indian classical music, the performer teases the listener by circling just above and below a stable tone, delaying the resolution as long as possible. When the resolution comes, members of the audience shake their heads and say, "Vah-vah!" Like life, music speeds up and slows down, it breathes, it has peaks and valleys of emotion, it engages our attention more or less strongly, it holds us then lets us go, and then picks us up again.

You can think of that stretched garage-door spring as containing potential energy—it *wants* to move; physicists also call this *stored* energy. When it starts to return to its original position, it is showing kinetic energy, the energy of movement. Similarly in music, composers and musicians create both potential and kinetic musical energy through a variety of means, principally involving pitch, duration, and timbre changes. But the "springiness" of music tension comes from our brains not from a physical object. No musical note is intrinsically or inherently "tense," rather, tension comes from *expectations* that our brains create based on stylistic norms for music, statistical properties of music, and the notes that we have just previously heard in the musical piece we're listening to. When we hear a note we didn't expect, or one that violates standard musical probabilities even in a small way, this is like pulling on the musical tension spring; our brains *want* the music to return to a more stable position. When we hear the first two notes of the chorus in "Over the Rainbow"—that big octave leap—it *feels* like someone has pulled a spring in our musical brains. The third note simply *has to* come down in pitch, and of course it does. In fact, the entire chorus of the song can be seen as an intricate and fabulous journey of trying to come to a relaxing resting point from that initial two-note tension. Joni Mitchell stretches the melodic spring several times in her song "Help Me," and spends the rest of the song allowing that spring to come nearly home, and—as in "Over the Rainbow"—not fully resolving the tension until the end of the song.

The tension in music motivates us to imagine musical scenarios that will come next—to form predictions. When our predictions come true, we feel rewarded and pat ourselves on the back. But we can learn even more when our predictions are not true, if events unfold in a way that is logical but is simply not one we would have thought of before ourselves. When a caveman friend showed another an easier way to find food, the second caveman recognized the value of learning, of expanding his repertoire of "right answers" as adaptive solutions to the problem of acquiring nourishment. Learning new things should feel good in our brains because it is usually adaptive.

Huron argues that music appropriates all of these four TRIP processes (plus another called "appraisal" that I've left out). He dissects the Beatles' song "She Loves You" to illustrate this. One of the most clichéd chord sequences in fifties pop music and doo-wop is what musicians call a I-vi-IV-V progression (in the key of G: G Major, E minor, C Major, D Major; sometimes a ii chord is substituted for the IV, that is, A minor, which has two-thirds of the notes in common with C Major). In the first section of the chorus, Huron notes, Lennon and McCartney throw in a C minor where we are expecting to hear a C Major. There is only one note different between a C minor and a C Major chord (E flat instead of E natural), but even non-musicians detect this instantly. The C minor doesn't last very long and then we are brought to the expected D Major—listeners reappraise the entire sequence, subconsciously of course, and realize that there exists a *plausible alternative* to the overlearned sequence they expected to hear. The listener, with the composers' help, has learned something new about the world.

If we think of musical sequences metaphorically as road maps, the point is clear. Caveman Og only knows one way to get to the watering hole and he follows that route every day. One day the route is blocked by a boulder that has fallen right in the middle. Fortunately, Og remembers what his friend Gluzunk showed him—a side trail that also goes to the watering hole. There is not only one way to get from point A to point B. Those of our ancestors who delighted in compiling information such as this—whether in the real world or metaphorically, artistically, *musically*—were those who were more prepared in the case of contingencies that interfered with attaining their goals.

The process of music listening thus involves tension, our reactions to that tension, our imagining and prediction of where the music is going to go next. All this can be seen as a preparatory activity for the sort of abstract thinking about the world that finding food, shelter, and mates—and escaping dangers—requires. For this to work, our ancestors had to enjoy playing this game of TRIP; they had to enjoy making predictions and then seeing if they were met or not. Remember that emotions (linked to motivation) are the way that our brains reward and punish us for actions that affect our fitness. By random mutation, some of our ancestors may have gotten a squirt of that feel-good hormone dopamine whenever they made successful predictions in the theater of their minds. This would cause them to want to do it again—to spend more time in thought, more time imagining scenarios, turning them over in their minds, playing the game of prediction and resolution over and over again. To the extent that such mental exercises conferred an advantage in the real world, in a relatively short amount of time this adaptation would permeate the population. To paraphrase Dennett, we don't sing and dance and get songs stuck in our heads because they are intrinsically attractive, memorable, or aesthetically beautiful. Rather, we have the relationship with music we do because those of our ancestors who found it enjoyable to *be* musical were those who were successful at passing on their genes.

Fundamentally, we have joy songs because moving around, dancing, exercising our bodies and minds is something that was adaptive in evolutionary history. Stretching, jumping, and using sound to communicate felt good because our brains—through natural selection—developed rewards for those behaviors. Joy songs today give us a jolt of good brain chemistry as a biological echo of the importance they held over thousands of years of evolution. "I define joy," Oprah Winfrey says, "as a sustained sense of well-being and internal peace—a connection to what matters." By being able to *celebrate* our good feelings, sense of well-being, and positive emotions, we were better equipped to share our emotional states with others, a key ingredient in being able to form societies and cooperative groups.

CHAPTER 4

Comfort

or "Before There Was Prozac, There Was You"

Eddie—the dishwasher at the pancake restaurant where I worked—lunged at my boss Victor with a kitchen knife. Victor fled, through the restaurant, just two steps ahead of him, knocking over a stack of high chairs and a few skinny teenage waitresses as he tried to get away. "I'm going to *kill* you!" Eddie shouted as Victor ran into the back, the Sunday morning patrons left agape. They had made two more complete circles through the restaurant and back room when Victor knocked over a tray of glasses. Eddie ran on over the broken glass without hesitation, passing by the griddle where I was cooking, and I grabbed his arm, pinning it momentarily to the counter. The knife fell out of his hand and straight down, lodging in his foot. The rest of the crew helped him pull the knife out while Victor made it to the parking lot and drove off. I went back to cooking pancakes and Eddie limped out the side door, and we never saw him again. All this over a song. And not just any song, but Tony Orlando and Dawn's "Tie a Yellow Ribbon Round the Ole Oak Tree." I understood

Eddie's frustration (though perhaps not his chosen cutlery-based response), because having music in the back room was what helped all of us get through the dreary workday. But it had to be the *right* music, and this was something that the ragtag bunch of us who'd ended up at the Newport, Oregon, Sambo's restaurant could never agree on. We all found comfort in music, and in fact it was this quality of music that had landed me there in Newport in the first place.

I dropped out of college to join a rock band, not because I found myself more interested in rock and roll than in calculus and physics (I didn't), but for my mental health. My school subjects were intellectually stimulating, but I hadn't made any friends my first year away from home and I was lonely. Music had always been a comfort to me during my largely solitary childhood. I found myself listening to music more and more during that first year away from home, and that made me want to *play* music more and more. The six songs that inspired me in 1975—that made me want to *become* a musician at least in those days—were:

1. "Autobahn" by Kraftwerk. These guys ushered in what is now regarded as an entirely new genre—techno (sometimes called *electronica*)—and like all great musicians, they made it seem easy. Combining long-form, classical themes with a kind of geeky interest in electronics and creating their own synthesizer sounds, they seemed to be just the kind of sciency, music-y nerds that I could aspire to being. The song from the album of the same name was a twenty-five-minute tour de force of drawn-out arpeggios and McCartney-ish shifts from major to minor tonality.
2. Beethoven's Sixth Symphony, performed by Herbert von Karajan and the Berlin Philharmonic. The development of the main themes and variations on them, the counterpoint

running through, are all laid out so transparently, so clearly for the listeners that it sounds as though this piece wrote itself. When listening to any one instrument, I felt that what the other instruments are doing at the same time is inevitable, obvious, that if only I could write a single melody, the other parts would present themselves to me automatically. Of course I knew that wasn't true, but the ease with which Beethoven navigated harmonic space inspires me still.

3. *Revolver* by the Beatles. (I can't single out one song from this album, but I listened to the album uninterrupted, hundreds of times, as though it were a song.) When this record first came out, I was only nine and not ready for it—I didn't discover it until college. The sense of play, fun, and camaraderie among the musicians is very noticeable, and the music still sounds fresh to me. Their songs took on a more sophisticated lyrical, melodic, and harmonic quality with this album than on previous albums by the group. They sounded like they were having such a good time!
4. "Through My Sails" by Neil Young with Crosby, Stills & Nash. The delicate vocal harmonies here still give me chills. Think of the Everly Brothers to the second power. The blend of their voices is gorgeous, comforting, and warm. It is usually Stills who performs songs with a solo lead vocal, joined at certain points by the lush background voices ("Suite: Judy Blue Eyes," "See the Changes," "Woodstock"), but here for the only time I've ever heard, Young pulls a Stills and delivers a few call-out lines, his voice enmeshed in the harmony the rest of the time. I'd always figured that Neil was too much of a loner, and too impatient to rehearse and practice, to work out the harmonic synchrony necessary to pull this off. Years later at a party, Graham Nash told me that this was

in fact true: That for this song, he, Crosby, and Stills had taken the tapes and painstakingly worked out their parts *after* Neil had recorded his. Neil told me not long afterward that he really doesn't like working over things, preferring the spontaneity of the moment. Neil even boasted to my friend Howie Klein—who at the time was president of Neil's record label, Reprise—that he had written *and* recorded an entire album in just two days.

5. "The Great Gig in the Sky" by Pink Floyd. This came from their concept album *Dark Side of the Moon,* the band's second effort (after *Meddle*) at playing with large-scale thematic connections among the tracks, more like a symphony than anything else that had ever been done in rock. The use of classical devices and narrative arc on the album got me thinking big, about merging classical and rock music. The singer, a guest vocalist named Clare Torry, delivers a chilling vocal performance—so much passion and sadness conveyed by the human voice that, even though there are no lyrics, the message seems so clear.
6. "Night and Day" by Stan Getz. As a sax player myself, I love listening to Coltrane, Cannonball Adderley, Wayne Shorter, and Charlie Parker, but Stan Getz has always held a special place for me because of his tone and his economical playing style. He was the first saxophonist whose parts I could get under my fingers; they just lay so easily on the instrument, and because of the hard rubber mouthpiece I used (and I suppose the shape of my mouth), I was able to emulate his tone more easily than the others'. I knew I could never play as fast as Bird, but I found comfort in knowing that a relatively slow player like Getz (Was he the B.B. King of the saxophone?) could still have a career.

My parents were naturally disappointed that I planned to drop out, unsure of what kind of future I could make for myself without a college degree. I told them about my plans to become a professional musician, but said I wasn't really sure how to go about it. They confessed that they didn't know either. I joined a series of rock bands that didn't get very far. We would collapse under the weight of our own incompetence, or we would simply find it impossible to get booked at clubs. After more than two years or so of such floundering, I came home to visit for my father's birthday in October. My father is a businessman who has always had a preternatural gift for problem-solving. His entrepreneurial spirit led him in his younger days to start his own accounting business, but he soon moved to a position at a large corporation. In his mind, there was an apt analogy here: Starting a band from scratch might be more difficult than joining a going concern, he told me. I realized that I had been playing with musicians who weren't much better than I, and in fact, I was often the best in these little groups that were forming. And I wasn't particularly good. That was no way to learn. If I was to become a professional, I had to be the *worst* member of the band. And so I set out to become the worst musician I could in a really good band! Or rather, I set out to become good enough that I could join a band of musicians who were just out of my league, who could help raise me up a notch in ability, who saw in me some potential that could be sculpted and molded.

My father gave me a book written by George Plimpton called *The X Factor* that described how people become experts in their respective fields. Plimpton points out that successful people have had far more failures than unsuccessful people. Of course this seems paradoxical. But the resolution is this: People who *eventually* become successful have had many, many failures along the way, and what distinguishes them from the rest of the population is that *they don't give up.* These leaders—corporate heads, expert chess players, actors, writers, athletes—look at failure differently than everyone else. First, when they fail, they don't assume that there is anything wrong with them ("I'm not good enough," "I suck"), nor do they figure that this is a permanent state ("I'm never going to get better," "I will always suck"). Rather, they look upon each failure as a necessary step toward reaching their ultimate goal. People who become successful see the progress toward a goal as involving a number of steps that will inevitably produce some minor setbacks. "This is something that I need to know in order to reach my goal," they say to themselves. "And until now, I didn't even know that I needed to know this. This setback is an opportunity to now go about acquiring the knowledge that is necessary to succeed."

In order to be the worst musician in a better band than the ones I'd been playing in, I knew I needed to practice more. But with the sort of menial jobs I could get without a college degree, all my income went toward the high rents in California, even with five people crammed into a three-bedroom flat (one person sleeping in the living room and one the dining room). I moved to Oregon, where I knew rents were cheap, so that I could spend most of my time practicing the guitar. I took a job as chef at Sambo's, a chain of pancake restaurants, and I could make my expenses in just two days a week of work—leaving plenty of time to practice the guitar. After six months of playing eight hours a day, I felt myself getting pretty good and I answered an ad at the local grocery store for a band that was looking for a lead guitarist, the Alsea River Band. This was a well-known band on the Oregon Coast, and they had actual gigs lined up several months in advance. They were led by a singer and songwriter named Étienne who was originally from Quebec. He may have been no older than forty-five, but that was ancient to me at the time, and his heavily creased face looked world-weary, battered, and he sang about love lost like he had lived every word. The band was a four-piece when I joined, Étienne playing rhythm guitar, a husband-and-wife duo on bass and keyboard, and a drummer. Étienne couldn't play lead, but he knew what he wanted to hear. He gave me cassette tapes of his songs as done by the band before they lost their lead player, and of some of his favorite music, by Hank Snow, George Jones, and Tammy Wynette. I had never played country music and I had no particular interest in becoming a country musician, but this was the only game in town, and they were good, and I mean *really good.*

So what if I didn't want to play country? My father said that in business (and music, he assumed) it was important to be flexible. Wouldn't I learn a lot about being a musician, about being in a band?

We rehearsed three nights a week in a trailer that the husband and wife were living in out in the forest, right on the Alsea River itself. No one had much work then; Oregon was in a recession. The drummer worked the counter at a car parts store, the bassist chopped wood, and his wife, the keyboard player, cleaned houses a few hours a week. Étienne had a straight job too, but no one talked about it. The first night I played with them was at the Waldport Lodge, which had seen better days. But it was a Friday night gig, a good night with a lot of people. The bartender introduced us as "an Oregon Coast institution—the Alsea River Band." Étienne took the microphone and launched into one of his signature tunes, the Shel Silverstein song "I Never Went to Bed with an Ugly Woman (But I Sure Woke Up with a Few)." I had a few little fills to play, nothing complicated. After a Hank Williams medley, we played "Mammas Don't Let Your Babies Grow Up to Be Cowboys." The crowd loved us and applauded boisterously after every number.

Then Étienne took off his cowboy hat, held it respectfully in one hand, and started talking to the crowd for the first time. He had never been farther south than Oregon, but he had perfected a sort of Memphis/Mississippi Delta accent. "We sher are happy y'all came out ta hear us tonaht, and we're a-gonna do are best to give y'all a little comfert at the end o' yer wee-uk." Étienne knew that most of the audience were unemployed locals, spending whatever they had managed to scrape together to come out to the lodge, to nurse cheap drinks as long as they would last, and to see an actual live band, a rare occurrence in these parts that would help them get through the drudgery of another two weeks or so without anything to do but look for work. "And tonaht yer all vury lucky, cuz we're featurin' somethin' rully special, our new guitarist, Dazzlin' Dan." He had named me that because he never could remember my last name, and even so, he couldn't pronounce it. I didn't feel particularly dazzlin'. For one thing, I had shown up wearing the same torn jeans and gravy-stained T-shirt that I wore to work, not being able to afford an actual performance wardrobe. Étienne would have none of that—he fished around in his duffel bag right before we went on and gave me one of his old jean shirts to wear. It was two sizes too big, but at least it was clean, and with the sleeves rolled up I looked almost stylish. Étienne wore a Western shirt with hand stitching on the shoulders and shiny, mother-of-pearl buttons. He had all the dazzle. "And Dazzlin' Dan," he continued, "is gonna show y'all a little of what five people kin dew tuh-gether when they's a-playin' good music." This was my cue to begin a little lead part that I had struggled over, a complicated fingerpicking pattern to open another of the group's signature tunes, "Poison Love" by Hank Snow. I played my bit passably, but the crowd was generous, maybe a little drunk, and applauded as Étienne slid into the first line: "Oh your poison love has stained the lifeblood in my heart and soul, dear . . ."

Standing onstage, surrounded by my new friends, I felt at home. I wished the song would never end. I hadn't lived long enough to have been done wrong by a woman, but Étienne and the crowd surely had. I started wondering about why the audience would be so happy to be reminded, lyrically, of infidelities and betrayals. It seemed as though there was comfort in numbers, camaraderie in a shared experience. And Étienne was a master at making it all seem okay. He was both a ladies' man and a man's man—women of all ages wanted to sleep with him, and men wanted to kick back and tell stories with him. Whatever he had done in his life, whatever he had been, his face and his voice were utterly without guile or deceit. I've known dozens of musicians like him, but only a few with his power to make everyone in a room forget about everything that was going on (or not going on) in his or her life, trading it all in for the extended reality of the four-minute song. "Yes," he seemed to be singing, "we've all been hurt, but that's part of life, and in the end, everything worked out all right and here we all be, together."

At Sambo's, the pancake restaurant, we had a little cassette player in the back room that housed the walk-in refrigerator, food preparation area, and dishwashing station. This is where I mixed the pancake batter and the waitresses refilled the Log Cabin brand syrup bottles with a cheaper, no-name substitute. Our dishwasher, Eddie, was a big, overweight, clumsy guy who had never finished the eighth grade. On my first day, several employees went out of their way to tell me to steer clear of him, that he was crazy. No one spoke to him. For reasons I never understood, Eddie took a liking to me, confiding his secrets, telling me his stories. His great ambition was to get a job at the Waldport Lodge of all places because they had a Hobart—a dishwashing *machine.* "Can you buh-leev it?" he asked me several times a week. "They's got a machine what cleans the dishes. Theys dishwasher just got to put them dishes in the machine. Set around waitin' until a li'l green light come on, and den he's a just gotta take them dishes out. Someday I'm wanna be the green light man. But youse gotta be in the *onion* to work there, and youse gotta be smart. And I ain't so smart."

Eddie was sensitive and kindhearted. He would bring me a muffin every day from the far side of the restaurant. Thinking that he had stolen it for me when no one was looking, he wrapped it surreptitiously in a paper napkin and presented it to me like a prize, singing, "Do you know the muffin man, the muffin man, the muffin man . . ." The ceremony was so elaborate I didn't have the heart to tell him that muffins were free to employees, and that as a matter of fact, as chef, I had cooked them in the first place.

The manager, Victor, had grown up in Las Vegas, and gone to Sambo's management training school at the corporate headquarters in Carpenteria, near Santa Barbara, California, and to his chagrin (he used words like *chagrin* quite often), he was sent to this small store on the Oregon Coast, far from the bright lights and big city that he felt were his due after all that training. It was Victor who hired me, on only his second day on the job. None of the locals liked him or trusted him. He drove a Mazda, the only foreign car in town. He wore white patent leather shoes with gold buckles and he used words that the coast-dwellers found strange. "I'm *chagrined* with you," he would say to a waitress who had fallen behind in her side work. He sized me up and confided that he'd only be working here until he could put a profit on the books, that he had hopes to be sent by Corporate to Salt Lake or Sacramento, or a "real" city, a place he could make a name for himself. If there was anyone the waitresses felt less comfortable with than Eddie the dishwasher, it was Victor.

The waitresses had made a mixtape for the little cassette player in the back and the six songs that they played most often back there were:

1. "My Guy" by the Supremes (a somewhat rare version compared to the Mary Wells version, which was the big hit)
2. "Stand By Your Man" by Tammy Wynette
3. "I Love a Rainy Night" by Eddie Rabbit ("Because he's so *cute,*" they said.)
4. "Nights on Broadway" by the Bee Gees
5. "Disco Duck" by Rick Dees
6. "The Morning After" by Maureen McGovern

Victor tolerated the girls' tape and had brought in one of his own with *his* favorite music, and when he was in the back—mostly at the beginning of the morning and the end of the day—he would take out whatever was playing and put in his music. Victor's top six songs were:

1. "Foreplay-Long Time" by Boston
2. "Two Tickets to Paradise" by Eddie Money
3. "Keep On Lovin' You" by REO Speedwagon
4. "Carry On Wayward Son" by Kansas
5. "Lady," by Styx
6. "We Built This City" by Jefferson Starship

When "Lady" came on, Victor would cock back the upper part of his body, play air guitar, grimace like a rock star, and mouth the words, looking now and then through squinted eyes all around to see if any of the waitresses were watching this profound display of his manliness.

Eddie was into heavy metal and had his own mixtape that one of his brothers made for him before being sent to prison for beating up three policemen with their own clubs. Eddie's tape was:

1. "Road Fever" by Foghat
2. "Warrior" by Wishbone Ash
3. "Paranoid" by Black Sabbath
4. "Running Wild" by Judas Priest
5. "Runnin' with the Devil" by Van Halen
6. "Tie a Yellow Ribbon Round the Ole Oak Tree" by Tony Orlando and Dawn

When any of Eddie's five heavy metal songs were playing, he would wash dishes with a vengeance, scrubbing them cleaner than the day they were new, and doing a little thing that sort of resembled a dance, with him shifting his considerable weight back and forth from one leg to another. He had worn holes in the thick rubber mats where he stood. But when "Tie a Yellow Ribbon" came on he would stand in place, his head hung over his shoulder, his arms limp at his sides, all the weight shifted to one leg. In Eddie's understanding of the world, the song was about a prisoner who would one day come home and nothing more.

What began as an uneasy truce between Eddie and Victor over control of the cassette player quickly degenerated. Victor didn't like Eddie's heavy metal and thought it was "inappropriate" for the restaurant. He'd substitute his tape, but as soon as he left the room, Eddie would put his tape back in, only to have it removed again by Victor. For more than a month, owing just to chance, Victor never walked in on the sixth song on Eddie's list, to see Eddie not working, idle and locked in his own thoughts. On this particular busy Sunday morning, Victor put in his mixtape and walked up to the front griddle with me to show me how to prepare a new menu item that corporate was introducing, a dish strikingly similar to one that our chief competitor, iHop, introduced two years later called the "Rooty Tooty Fresh 'n' Fruity Breakfast Plate." This proto Rooty Tooty Fresh 'n' Fruity plate consisted of a pancake with two orange slices and a fried egg on top, and a sausage and slice of bacon below arranged to look like a mouth on the pancake's face. Victor's tape ended and after a few minutes of musical silence, I saw through the door into the back as Eddie opened the cassette machine. Victor's tape flew out from the spring-loaded mechanism, and onto the floor into a pile of sudsy water. Eddie ignored it and put his tape in. He hadn't rewound from the last time, and it was cued up to the Tony Orlando song.

"I'm coming home I've done my time," the singer began. Victor made a face at me and stormed into the back. "Who put this crap music on?" he demanded. He whipped around and stared at one of the waitresses, Tiffany, who was stuffing napkin holders. "Who?" he asked again. Tiffany shrugged. Victor then noticed Eddie, standing motionless with his head hung low over his sink. "Eddie! Why aren't you working?!" Eddie just stood there, silently mouthing the words. Victor paused for a second and then something clicked. Hiking up his pants, he leaned forward. "Oh, no! Don't tell me!" he said in a mocking tone. "*You?* You put on this crap sissy-boy music? Ha ha ha ha!" Victor didn't just laugh, but held his stomach, like a caricature of someone laughing, and pounded the side of his leg. "I should have known that stupid Eddie would put on this stupid music," he stretched his arms out wide, turning his body halfway around, announcing to the nearly empty room.

"Don't call me stupid," Eddie said, still staring down at the soapy sink, his back to the rest of us. "I ain't smart. But I ain't stupid neither."

Victor wasn't listening. "Ohhhhhhh," he said with an exaggerated downward glide, "tough dishwasher man likes to listen to little baby music. Baby music!" He laughed some more and pointed his finger at Eddie.

"Let it go," Tiffany said softly.

Victor grabbed Eddie by the arm, but Victor's small hands couldn't close around Eddie's large forearm. "Turn around when I'm talking to you, stupid!"

Eddie turned around and he was crying. "The song—the song's about my *brother,*" Eddie said.

"Your *brother*?" Victor mocked him.

"Please," Tiffany said, a little louder this time, "Victor, *please*—just let it *go.*"

"Your brother? I've got news for you, *stupid*. Your brother isn't coming home." Victor taunted him. "Didn't you read in the paper last week?" Victor knew that Eddie couldn't read. "One of those cops *died*. Your brother is staying in prison for a long time. He'll never be tyin' no yellow ribbons around no tree, oak or nutherwise; the only thing he'll be tying is his own hangin' rope. Do you hear me, you big stupid lug? He's *never* coming home."

Eddie furiously grabbed one of the soapy knives out of the washbasin and drew his arm back, tears rolling down his cheeks. Victor sprang backward, smashing his own mixtape where it lay on the floor, and then flew through the back to the griddle area, where

I was putting the finishing orange slices on a Proto Rooty Tooty Fresh 'n' Fruity Breakfast Plate. Eddie followed right behind, the two of them yelling and screaming as our Sunday morning patrons looked on, half-astonished and half-scared out of their wits, while the chorus played jauntily on, "Tie a yellow ribbon round the ole oak tree/It's been three long years/Do you still want me . . ."

We never saw Eddie again at the restaurant, and I read a few months later that his brother was killed in a knife fight in the state penitentiary. I did see him on the street a few times, standing in line at the unemployment office, and stopped to say hi. He called me "Dan Dan the Muffin Man." We never spoke of the knife incident. Victor instituted a new rule that only he was to control the music in the back room. He hired another dishwasher, a man who was legally blind, whom he would taunt endlessly. "There's a *spot* on this plate," Victor would say, holding up a perfectly clean plate. "You'll have to do it again. Be more careful—can't you *feel* the spots? I thought you blind cripples had highly developed extrasensory perceptions. Can't you *feel* the spots?"

In many of the places I've worked, music has been there as a soundtrack to help the employees get through their day. Of course there is no one song that everyone likes and it can be a challenge to please everyone, but when the right balance is struck, music helps to break the monotony, to comfort us through boring or stressful tasks. Many surgeons I know listen to music in the operating room—even brain surgeons! When I worked as an auto mechanic, the radio in the garage was going nonstop, tuned to the hit rock station. In my laboratory at McGill, where eight to ten people all work in a large room together, each computer workstation is equipped with its own stereo speakers and subwoofer. If the different music starts to compete, each computer station also has a set of headphones, and the students are typically found listening to their music as they perform their statistical analyses or analyze brain images. We all hear music in buses, train stations, dentists' offices, elevators, Wal-Mart, when we're on hold. The purpose of this is ostensibly to comfort.

Mothers from every culture sing to their infants, and have done so throughout time as far as we know. Singing can soothe and comfort infants in ways that other actions cannot, and this is in part because of how different auditory stimulation is from other senses. Sound can be transmitted in the dark, even when the baby's eyes are closed. Auditory signals feel as though they come from inside our heads, unlike visual signals, which appear to be "out there" in the world. Before the infant's visual apparatus is fully formed—before it can make out the difference between its mother and other adults—the auditory system is capable of recognizing the consistent timbre of its mother's voice. Why is it that mothers instinctively sing rather than speak, and why is it that babies find song especially comforting? We don't have the answers to this, but neurobiology shows that music—but not speech—activates areas of the human brain that are very ancient, structures we have in common with all mammals, including the cerebellum, brain stem, and pons. Song has repetition built into it—of rhythms, melodic motifs—and this repetition gives song an element of predictability that speech lacks. This predictability can be soothing.

The lullaby is the classic song of comfort. Most lullabies that we know of share structural similarities, according to my friend Jonathan Berger. Jonathan is a very highly respected composer and music cognition researcher, as befits his position as a tenured professor at Stanford. We met in his office on the Stanford campus to talk about *Six Songs,* then wandered down to the Stanford Bookstore, where, surrounded by musical scores in their extensive collection, we continued over lattes.

"I think lullabies are, in a way, arguably in a separate class because (a) they're functional; they're for calming someone else. They're not for calming you. And (b) they have a formulaic pattern. David Huron mentions this, where it's a big leap and then a slow descent down and the idea is that you grab the attention and then you decrease arousal. And so there's sort of a melodic pattern to lullabies that puts them in a class by themselves.

"And almost any lullaby fits that melodic pattern. In my undergraduate music cognition class I ask—because it's a very international group of students—'Sing your first lullaby that comes to mind' and they all fit that rule. There is none that I've come across that doesn't fit. [Sings Brahms' lullaby: da da dee, da da dee, da da DEE da da da dah] There it is, on the ninth note, the large leap. And then stepwise motion from there."

Music theorist Ian Cross disagrees that lullabies are only for comforting the infant. "First-time mothers experience a great deal of uncertainty and apprehension about their newborns. 'What do I do with *this*?'" Singing mutually calms the mother and child. Because it requires regular, rhythmic breathing, it can serve as a kind of meditation for the mother. The slow, steady rhythms of singing lullabies can stabilize respiration, and also heart rate, lower the pulse, and cause muscle relaxation.

Another and perhaps not obvious form of comforting music is music made for the disaffected and disenfranchised. Teenagers who feel misunderstood, cut off, and alone find allies in lyricists who sing of similar alienation. In affluent societies around the world, so many teenagers feel as though they don't fit in, that they're not among the cool; they feel lonesome and alone.

These function as bonding and friendship songs, for sure, and simultaneously as comfort songs. In the seventies, some of us listened to musicians who sang about things that were not discussed—free sex, smoking cigarettes or marijuana out in back of the school (think Brownsville Station and "Smokin' in the Boy's Room" or the Animals and "Tobacco Road"). The implicit message of these songs was "You're one of us—you're not alone—the things you think and feel are normal." Janis Ian's "At Seventeen" addressed the millions of teenage girls (and boys) who felt they didn't fit in because they were not attractive enough:

To those of us who knew the pain
Of valentines that never came
And those whose names were never called
When choosing sides for basketball
It was long ago and far away
The world was younger than today
And dreams were all they gave for free
To ugly duckling girls like me.

In the eighties and nineties, Michael Stipe with R.E.M. and Morrissey (first with the Smiths and then as a solo artist) reached millions of listeners with their songs of depression, alienation, and detachment.

Today when high schoolers feel misunderstood they listen to hip-hop and rap and lyrics such as "Gangsta's Paradise" by Coolio:

I'm living life do or die, what can I say?
I'm twenty-three now but will I live to see twenty-four
The way things is going I don't know.

Some comfort songs are refrains, intended to calm us in the face of danger, or soothe us when facing death (either our own, or the death of a close one). In "Death Is Not the End," Bob Dylan creates a simultaneously anthemic and anesthetic refrain: "When you're sad and lonely and you haven't got a friend, just remember that death is not the end . . . When the cities are on fire, with the burning flesh of men, just remember that death is not the end." As with many songs, the lyrical intent is ambiguous. Is Dylan singing to someone who has just lost a friend, telling her that the friend is not really dead? Or is he suggesting the recipient of the song might consider suicide? Either way, the message that death is just a portal, not the end, and you will live on afterward, is more comforting than the alternative, that death ends everything definitively.

David Byrne described three songs that he reaches for when he feels he needs comfort, or in his words, "consoling": "I Don't Wanna Talk About It Now," "Michelangelo," and "Boulder to Birmingham," all written by Emmylou Harris, whose voice Rodney Crowell describes as "equal parts siren and Earth angel—the embodiment of the feminine, the voice you'd want to take to the oft mentioned deserted island." ("Boulder" was cowritten by Bill Danoff, who wrote "Take Me Home, Country Roads," the John Denver hit, and "Afternoon Delight," a hit for the Starland Vocal Band.)

I don't want to hear a love song, I got on this airplane just to fly
And I know there's life below
But all that it can show me is the prairie and the sky

"'Boulder to Birmingham' is the one she did with Gram Parsons. Somebody beautifully pouring out some pain but not just screaming out like they hit their thumb with a hammer. Pain, but in a much more slow, heartfelt way."

I asked if he ever picked up the guitar and sang one of his *own* songs for consolation, if they had that effect on him.

"Once in a while," he said. "There's a few—usually more of the recent ones. There are a few that I occasionally play, that I enjoy. The singing is kind of consoling, or cathartic, or soothing to me; which is something that I always wanted to be able to do: to write a song that would be a tool that I could use for myself the way I've been able to use other people's songs. There's two from my album *Look into the Eyeball:* 'The Revolution' and 'The Great Intoxication.'"

The Revolution

Amplifiers & old guitars
Country music sung in bars
& when she sings the revolution's near

Beauty holds the microphone
& watches as we stumble home
& she can see the revolution now

Dirt & fish & trees & houses
Smoke & hands up women's blouses
Not like I expected it would be

Bubbles pop in every size
It's analyzed & criticized
& beauty knows that it is almost here

Beauty goes to her address
She shuts the door and climbs the stairs
& when she sleeps the revolution grows

Beauty rests on mattress strings
Wearing just her underthings
& when she wakes the revolution's here
& when she wakes the revolution's here

After the attacks of September 11, 2001, Americans were in need of comfort. To a majority of us, the unthinkable, the unanticipatable had occurred with the sudden, coordinated set of surprise attacks on U.S. soil. It wasn't just our national pride that was injured, but our very sense of safety and security. Many commentators noted that interviews taken with Americans on the street over the few weeks immediately following the attacks were notable for this profound sense of injury, and also for a relative lack of aggressive feelings of wanting to retaliate—those feelings of a militaristic nature came only later, and (arguably) as a result of political rhetoric from the White House. During the initial aftermath, radio and television stations, train stations and bus depots, and many public places began to pipe music to Americans. And what did they play? Not the battle-tinged refrains of "The Star-Spangled Banner," but a song written by an immigrant in 1918, near the end of World War I, Irving Berlin's "God Bless America." "That song spontaneously became our de facto national anthem," says Rabbi Yosef Kanefsky, a spiritual leader in Los Angeles, "at a time when people were looking for something with which to express themselves, and to bond. It's amazing—the capacity of that simple melody to unite the country in a way that was both comforting and brought strength—it crossed all divides."

My friend Amy was diagnosed with a brain tumor a few months ago and is now undergoing radiation therapy. Every day she has to report to the hospital and lie perfectly still for an hour, with a Hannibal Lecter–style titanium mask bolted to her head to prevent even the slightest movement that could cause the proton density beams to miss their mark. It is a very uncomfortable and frightening experience for her. The neurosurgeon who is treating her told her to bring in music every day to the treatment sessions because he knew, based on published research, it helps to relieve anxiety and reduce the painful effects of the procedure. Amy brought in Sting's *The Dream of the Blue Turtles* for her first session, and *Nothing Like the Sun* for the second one. I doubt Sting ever imagined his masterpieces being used in this context, but it has transformed for Amy what could have been a nausea-filled, adrenaline-toxic ordeal into a tolerable if not somewhat aesthetic experience.

Country music lyrics often tell of a love gone bad, or of Hank Williams's now iconic "cheatin' heart." So much of recovery is knowing we're not alone and that we're understood. And good music, like good poetry, can elevate a story to give it a sense of the universal, of something larger than we or our own problems are. Art can move us so because it helps to connect us to higher truths, to a sense of being part of a global community—in short, to not being alone. And that is what comfort songs are all about.

While I was dining with Joni Mitchell at an outdoor restaurant once, two women in their late forties approached us, recognizing her. "We just wanted to thank you," they said, apologizing for interrupting her meal. "We had a really hard time getting through our twenties. This was the 1970s," they explained. "We listened to your album *Blue* and it made us feel better. Before Prozac there was *you*!"

When we are sad, many of us turn to sad music. Why would

that be? On the surface of things, you might expect that sad people would be uplifted by *happy* music. But this is not what research shows. Prolactin, a tranquilizing hormone, is released when we're sad. Sorrow does have an evolutionary purpose, which is to help us conserve energy and reorient our priorities for the future after a traumatic event. Prolactin is released after orgasm, after birth, and during lactation in females. A chemical analysis of tears reveals that prolactin is not always present in tears—it is not released in tears of lubrication of the eye, or when the eye is irritated, or in tears of joy; it is only released in tears of sorrow. David Huron suggests that sad music allows us to "trick" our brain into releasing prolactin in response to the safe or imaginary sorrow induced by the music, and the prolactin then turns around our mood.

And aside from the neurochemical story, there is a more psychological or behavioral explanation for why we find sad music consoling. When people feel sad or suffer from clinical depression, they often feel alone, cut off from other people. They feel as though no one understands them. Happy music can be especially irritating because it makes them feel even *more* alone, less understood. I now know that my boss at Sambo's, Victor, was probably suffering from clinical depression, and took out his sense of powerlessness on those who were weaker then he. The upbeat, happy song of Tony Orlando and Dawn was enough to make him snap in this condition. When we are sad and hear a sad song, we typically find it comforting. "Basically, there are now *two* of you at the edge of the cliff," says Cambridge University music professor Ian Cross. "This person understands me. This person knows what I feel like." That connection—even to a stranger—helps the process of recovery, for so much of getting better seems to rely on feeling understood—one of the reasons why talk therapy is so successful in cases of depression. In addition, the depressed person reasons, this person who went through what I went through lived through it; he recovered and can now talk about it. Moreover, the singer turned that experience into a beautiful work of art.

The blues may be the ultimate comfort song in Western society during the last hundred years. The "blues" technically refers to a type of chord progression, in its simplest form what musicians call I-IV-V7 (pronounced "one, four, five-seven") and the many variations and reharmonizations of this basic progression, typically done in twelve- or sixteen-bar phrases (hence the term "twelve-bar-blues"). The lyrical content to this chord progression can be anything, from the Beach Boys praising the beauty of their local beaches and babes ("California Girls"), Chuck Berry paying homage to an especially skillful guitarist ("Johnny B. Goode"), or Steely Dan exploring Eastern enlightenment through the Buddha ("Bodhisattva"). But the prototypical lyric is about someone who has had hard luck, been done wrong by life and circumstance, and this is what makes the songs comforting—the idea described above that sad people are so often made to feel better by sad music.

"Going Down Slow" written by St. Louis Jimmy Oden (and performed by Howlin' Wolf, the Animals, Eric Clapton, Led Zeppelin, and Jeff Beck with Tom Jones, among many others) is one of the thousands and thousands of amazing blues songs that have been a cultural legacy of black America throughout the twentieth century and beyond. It is the song of a dying man, looking back on his life, asking for his mother to come to his deathbed. It is heart-wrenching and bittersweet. The Jeff Beck version is one of the most powerful blues performances I've ever heard. Tom Jones trades in his normal sexy, confident swagger and inhabits the role of the luckless and life-scarred vagabond, his voice all but unrecognizable, dripping with despair and misery. Beck—widely regarded as among the top five guitar players alive today—plays what may be one of the most powerfully emotional electric guitar solos ever recorded. I played this last week for Sandy Pearlman, producer of Blue Öyster Cult and the Clash. Sitting in my car as we sped down the Trans-Canada Highway, he closed his eyes and smiled with every new note that Beck played. After the first vocal line, as Beck began to play his "call-and-response" guitar fills, Pearlman beamed and said, "Now *there* is a guy who understands *everything* about the cognitive neuroscience of emotion and music! He *knows* just what to play to put goose bumps on the hair that is standing up on the back of my neck!"

A sad song brings us through stages of feeling understood, feeling less alone in the world, hopeful that if someone else recovered so will we, and we feel ultimately inspired that the sad experience led to something aesthetically pleasing. For people who were currently sad at the Waldport Lodge that night, Étienne showed them hope. For those who were over their sadness, Étienne reminded them of how far they had come, having successfully traded despair for at least a temporary peace until the next sad episode might appear.

CHAPTER 5

Knowledge

or "I Need to Know"

I came to academia late—I didn't even get my B.A. until I was well into my thirties—but I had known Ian Cross's name many years before we first met, principally from two important books he co-edited on musical structure and from articles on the cognitive representations of musical form. The field of music cognition is relatively small; there are probably only two hundred and fifty people in the world who would consider it their specialty. Contrast this with a field like neuroscience, which draws 30,000 attendees to its annual conference in the United States alone. Most university psychology and music departments don't have *anyone* doing music cognition, and those that do rarely have more than one. This makes the annual meetings of the three main societies (the North American, the European, and the Pan-Asian) a big deal—it is an opportunity for people in the field to meet and learn about the latest findings, to resolve scientific controversies, and to schmooze.

In graduate school I completed some new work on absolute

pitch, and I presented it at the European conference one year to get feedback on it before sending it out for publication. Research scientists are famously intolerant of logical arguments or experimental designs that are flawed, or young researchers who make claims that are not fully supported by the data. This is trial by fire for a student, but there is no better training. Attendees may shoot your work full of holes, but in the end if you are able to plug up the holes, the paper becomes stronger. As my doctoral advisor, Mike Posner, counseled me, it's better to know *before* the paper is published if there are any flaws; printing retractions is embarrassing and career-stopping.

On the first morning of the conference I came down for breakfast and sat at a large round Formica table with some other students whom I had met on the bus ride in—two of them were Ian's students. They were a friendly bunch, curious to know about my background and about the paper I was going to give, what graduate school was like in America, if I had ever met a Hollywood actor (of *course*! America is crawling with them!), what bands I liked to listen to. Ian came in later, wearing a sharply pressed suit, and, to my surprise, sat down at the table with us students. I had assumed that meals at the conference would be similar to my family's Thanksgiving dinners, where there were separate tables for the adults and the children. (And where, no matter how old I became, I still was relegated to the children's table because my parents' generation were still filling up the adult table. Last Thanksgiving several of us second-generation family members teetered uncomfortably on half-sized, low chairs at a correspondingly low table—the same chairs we've been sitting in for forty years.)

Ian introduced himself to everyone and struck up conversations with us all. He had himself done some work on absolute pitch and so was looking forward to my talk, he said. How often do professors go out of their way to get to know students like this? (Not often!) My first impression of Ian was of his generosity, and his indifference to social rank. We had another thing besides absolute pitch research in common: Ian was a guitarist, and although he played classical guitar and I played blues, we both had spent our lives listening to each other's favorite music and understood it. That night after dinner, some of the attendees took turns taking the stage to play music for the group. Ian insisted that I borrow his classical guitar to play something for everyone, and so I sang and played a song I had just taught myself that month, Stevie Ray Vaughan's "Pride and Joy." Mine was the only nonclassical contribution to the evening, but everyone seemed to be having a good time. To this day, I still run into people from that conference who remember me as "that guy who played Stevie Ray Vaughan." I've known Ian fifteen years now and he continues to astonish me with his clarity of thought, his scholarship, and his insights into music cognition. Over the past ten years, Ian has written a number of papers on the evolutionary origins of music, and they have become influential in the field.

We were both invited to deliver addresses at the International Music Meets Medicine conference hosted by the Gyllenberg Foundation in Espoo, just outside of Helsinki, in summer 2007. The days were long, with twenty-one hours of sunlight. I had never been that far north before, and Ian and I both noticed how different the color spectrum of the sun appeared—everything seemed a bit more yellow to us. We walked around the conference center taking in the new-to-us vegetation. The trees looked similar to trees we knew from our respective childhood homes, Scotland and California, but we noticed subtle differences in bark patterns, in the color of leaves, and—on the conifers—needles. We speculated about whether these were different species of trees than the ones we grew up with, or simply genetic variations that had adapted to fluctuations in the amount of sunlight they would be exposed to, varying seasonally from two to twenty-two hours a day. The fauna were also different—as we sat at a picnic table and talked about musical origins, our conversation paused frequently when we heard an unfamiliar bird, which prompted a spontaneous joint effort to try to see it and identify it.

We were also distracted by the calls of frogs in the pond next to our picnic table, and eventually were able to see some small, smooth-skinned brown ones that were unlike any we had seen in our own countries. The frogs and toads that Ian and I knew from childhood croaked together in a kind of amphibious symphonic *tutti*. The Finnish frogs—or at least those in our pond in Espoo—used what biologists call *antiphonal calling*. This, we read later in a field guide, is to maximize the chance that a female frog will be able to find her favored male. Frogs choose their mates largely on the basis of sound. A female frog can be swept off her legs by just the sound of a suitably enticing male frog call played over a loudspeaker, even trying to mate with a stuffed replica of a male. The reason that most frogs synchronize their calls is that it makes it more difficult for predators to locate them; the antiphonal calling in Espoo must have developed as a competitive advantage in mate attraction for those males who employed it, somewhat at their own increased risk of being eaten by predators.

Ian began our conversation on musical origins by considering what the requirements might be for *any* system of animal or human communication. "Clearly," he said, "the survival prospects of individuals and groups are enhanced by a capacity to communicate certain information about states of affairs in the physical world, and in the social world that concerns the organism. Even more so by the ability to organize action in response to those states of affairs," actions such as running, hiding, fighting, cooperating, and sharing.

"Of course," I interjected, "as David Huron would say, survival is only enhanced by sorting out fact from fiction, meaning that the organism requires the ability to detect liars, manipulators, and exaggerators." This is exactly the argument that Huron and others have made for the value of music over language. This is a bold and controversial notion that is gaining favor among researchers. What you want for a communication medium is one in which honesty can be readily detected, what ethologists call an *honest signal.* For a number of reasons, it appears that it is more difficult to fake sincerity in music than in spoken language. Perhaps this is simply because music and brains co-evolved precisely to preserve this property, perhaps because music by its nature is less concerned with facts and more concerned with feelings (and perhaps feelings are harder to fake than supposed facts are). Music's direct and preferential influence on emotional centers of the brain and on neurochemical levels supports this view.

Ian continued that the ideal communication system would allow individuals to communicate knowledge about current conditions such as the availability and locations of resources, to make possible their sharing; perceptions of dangers would need to be identified and appropriate actions coordinated; finally social relationships would need to be articulated and sustained. Why is music necessary and even better than language for such tasks? I think it is because music, especially rhythmic, patterned music of the kind we typically associate with songs, provides a more powerful mnemonic force for encoding knowledge, vital and shared information that entire societies need to know, teachings that are handed down by parents to their children and that children can easily memorize. I believe that this is such a fundamentally important *function* of music that it may even have been the root of the first song (notwithstanding Sting's and Rodney Crowell's Chapter 3 lobbying for joy songs as the primeval musical form).

Imagine an early ancestor of ours, maybe a hundred thousand years ago, standing above a river where there is a gathering of crocodiles. Another early human is near them, and our ancestor hears one of the crocs make a certain noise before chasing and then devouring the nearby human. Our ancestor has learned that this noise is the signal that the croc is about to attack. Due to a random, unexplained mutation, his frontal lobe is a bit larger than anyone else's. He has a greater than normal capacity to reason and to communicate; in particular, he has a perspective-taking ability, albeit a rudimentary one, an ability to imagine what other people are thinking. He realizes that this knowledge that he has is not knowledge that his children have. They are precious to him. He wants to warn them.

He runs home and (like all other humans at that time) has no language, but he feels it necessary to communicate to his children the danger he just witnessed. He doesn't want to bring them to the scene; too dangerous—they could become dessert. He needs to communicate the danger symbolically. He imitates a croc. He gestures, wiggles on the ground, uses his body to make the motions of a croc. He brings his arms and hands together to mimic the jaws opening and closing, then makes the noise. This type of symbolic gesture may be practiced for thousands of years until a further refinement is introduced, precipitated by an even more enlarged frontal lobe. The young children don't necessarily pay attention to this vital message; they are playing, babbling, making noises, laughing, moving, wiggling. The father incorporates their behaviors into his in order to attract their attention. He laughs, moves, makes funny noises. He surrounds the important message about the crocodile and the crocodile's noise with an attention-getting dance, accompanied by pitched and rhythmic vocalizations. The first song is born, and it is born to simultaneously educate, grab the attention of, and entertain children. Today children are remarkably attuned to the music around them; they rock and sway to music in their environment, and within their first two years develop their own preferences for music. They are also attuned to the music within them, genetically encoded, entering a period of musical babbling often even before their linguistic babbling begins.

Children's penchant for music seems to begin in infancy. By seven months, infants can remember music for as long as two weeks and can distinguish particular strains of Mozart they've heard versus very similar ones they haven't, suggesting an innate—and evolutionary—basis for music perception and memory. And as Sandra Trehub has shown, mother-infant vocal interactions exhibit striking similarities across a wide range of cultures. These interactions tend to be musical, with wide pitch ranges, repetitive rhythms, and with clear emotional and instructive (knowledge-giving) content. Instinctively, mothers and infants co-regulate affect through these interactions, mothers reassuring their infants that they are nearby and attending to them. Mothers also use these musiclike vocalizations to direct their infants' attention to important perceptual features in the immediate environment.

David Huron suggests that the first song may have been more related to *pride* than fear, as in my crocodile scenario. "Imagine," he says, "you've gone out on the hunt and you've come back—a group of you—and you want to share what happened with the others who weren't there. And yet, you want to give them an aesthetic experience of it; you don't want to report the way a bee would, 'This is where the meat is.' You want to report in an artistic fashion; in a highly stylized form to convey the sense of danger, your difficulties, your ultimate accomplishment." This may have begun as pantomime and evolved into something recognizable as music-dance.

Alternatively, Ian Cross suggests, the first song may have grown out of a children's chanting game, a turn-taking game such as "Patty-Cake" that helped them to coordinate their movements with those of another person.

In all cultures that have a number system, children have counting songs, rhyming ditties, to help them learn their number line by rote. In our culture these can be partly sung and partly spoken, and they typically do double duty to train motor coordination as in jumping rope songs:

Down by the river, down by the sea,
Johnny broke a bottle and blamed it on me.
I told ma, ma told pa,
Johnny got a spanking so ha ha ha.
How many spankings did Johnny get?
1, 2, 3 . . . [keep counting until the jumper makes a mistake]

or

Cinderella, dressed in yella
went upstairs to kiss a fella
made a mistake
and kissed a snake
how many doctors
did it take?
1, 2, 3 . . . [keep counting until the jumper makes a mistake]

By the age of three, many children are already making up their own songs, or versions of songs they've been taught, generating variations on the heard melodic/rhythmic patterns of their culture in much the same way they generate variations of speech patterns. This sort of spontaneous experimentation suggests that the predisposition toward melody and rhythm variation is hardwired in the brain; it may have been necessary to our ancestors, contributing to reproductive fitness.

Ian and I continued to talk by the Espoo pond. "Ultimately," Ian continued, "music developed as a 'communicative medium optimally adapted for the management of social uncertainty.'" Whether music preceded or followed language is not the point, Ian argued, because for tens of thousands of years *both* would have existed, and evolution, the brain, and culture would have accommodated to both.

"The very thing that music lacks—external referents—makes it optimal for situations of uncertainty," Ian continued. "When social situations are difficult, confrontational—such as encounters with strangers, changes in social affiliations, disputed courses of action—the fact that language so unambiguously denotes individual feelings, attitudes, and intentions can tip situations into dangerous physical conflict. Language can become a social liability. But let's imagine the possibility of access to a parallel system of communication, one that by its very nature tends to promote a sense of affiliation, unity, bonding. *And . . .*" Ian paused, his eyes reflecting the water of the pond, "one that conveys an *honest signal*—a window into the true emotional and motivational state of the communicator."

And as a signal of emotion, there may be none better than music. Consistently, across all cultures we know of, music induces, evokes, incites, and conveys emotion. This is especially true of the music in traditional societies. And in laboratories, music is probably the most reliable (nonpharmaceutical) agent we have for mood induction. If music and mood/emotion are that closely tied, there must be an evolutionary explanation.

One evolutionary explanation for the relationship between music and emotion comes from the awareness that emotion is intimately related to motivation in humans and animals. So in looking for the connection, we might first ask: How could music have served as a motivator among nonhuman animals? Brains coevolved with the world and have incorporated certain physical regularities and principles of the physical world. One of those principles is that *larger objects,* because of their increased mass, tend to make sounds of a lower pitch when they impact with the earth, or when they are struck (because in the latter case, their resonant frequency is lower as a function of their larger size).

The ancestral mouse that learned to pay attention to low-pitched sounds would have avoided being stepped on by elephants. In fact, very few of those that lacked this ability would have ended up being ancestors to any future mice, because they would have gotten stepped on. A sensitivity to certain frequencies, and to intensity and rhythm, would have been important. You don't want to be *too* sensitive and startle to everything, or you end up staying in your mousehole all the time and you never get out to acquire food or a mouse mate. You need to startle to all the right things but only those. At the same time, if low-frequency signals indicate large size, some mice might have stumbled upon the fact that if *they* made low-pitched sounds with their throats and mouths, it might serve to intimidate other mice (if not elephants).

It may be a long way from frequency sensitivity in mice to music in man, but it is a robust beginning. Thousands of small adaptations would have helped different species to find their ecological niche. It only takes one line of mutations in a single family tree to combine pitch selectivity with rhythmic sensitivity, and the foundations of the musical brain are there, waiting to be exploited when an enlarged prefrontal cortex figures out what to do with all that auditory discriminability. Before looking at how these brain developments may have occurred, however (which I'll unpack in Chapter 7), I'd like to look more closely at just what knowledge songs really are and how they work.

Today, music is produced by few and consumed by many. But this is a situation of such historical and cultural rarity that it should hardly be considered. The dominant mode of musicality throughout the world and throughout history has been communal and participatory. We've seen the change even in a few generations. One hundred years ago, families would gather around after supper and sing and play music together to pass the time. In her memoir of 1880s Manhattan, Paula Robison writes:

> Vicarious musical pleasure by radio and phonograph, while it encourages listening to good music, seems to put a damper on musical self-expression. [In our childhood] we sang more. Children sang at school and in their play. Folks sang as they worked, indoors and out. Even drunks do not sing in the streets and buses as entertainingly as in [those] days.

We see echoes of our shared history today in summer camps and on school buses. Knowledge songs may well have been the first, and David Huron observes that the flavor and sense of them is preserved here in North America in what he calls "yellow school bus songs." These are songs such as "99 Bottles of Beer on the Wall" and "The Ants Go Marching" and even "The Wheels on the

Bus Go Round and Round." Songs like this are primarily teaching songs. "99 Bottles" and "The Ants" teach children to count. "The Wheels on the Bus" helps to construct and reinforce the physical and social order of the environment, encoding the perceptions into age-appropriate schemata: The baby in the bus cries, the wheels go round and round, the wipers go whoosh-whoosh, and so on. These songs simultaneously teach children things they need to know about the world and about musical forms and structure.

Another class of songs sung or chanted by children all over the world is selection or counting-out rhymes, the most famous of which in North America is probably this one:

Eenie, Meenie, Miney, Moe
Catch a tiger by the toe
If he hollers, let him go
Eenie, Meenie, Miney, Moe
[At this point, regional variations kick in; one researcher found several dozen different endings. The one I learned went like this:]
My mother told me to pick the very best one and you are not it.
[The person pointed to is "out," and the game continues until only one player is "in." That person then is "it" in whatever activity is being selected for.]

The interesting thing about such rhymes is that they are passed on almost entirely by oral tradition. No child reads the rhyme in a book, and typically children learn them from *other* children, not from adults. The rhythmic aspects of the songs and the vocal-motor coordination required to point and recite them effectively are practice for more adult activities. Children who are part of such a circle of counting out are very vigilant about violations of pointing or counting, making the reciter start over if there is even the slightest mistake. The game is socially reinforced and serves as preparation for more sophisticated songs that children inevitably learn, and that carry more important meaning. Across all of North America, the fixedness and similarity of different versions is extraordinary.

Many children's songs also help to train memory, and although the songs themselves do not impart knowledge, they are the juvenile precursors of epics and ballads that do. "I Know an Old Lady Who Swallowed a Fly," or "An Only Kid, An Only Kid" are examples of songs that continue on, with each verse invoking earlier verses in an interconnected narrative, so that by the end of the song the memory load is quite high. Young children typically remember isolated phrases and try to emulate older children who can make it all the way (or nearly all the way) through the entire story. Vivid imagery and animals—both things that appeal to children's developing imaginations—help to preserve the concepts of these songs, and the children may learn the words as secondary, or subsidiary, to their mental images of the story. The most effective of these songs additionally use poetic devices—rhyme, alliteration, and assonance—to help constrain the possible words and give children a jump start in memorizing them. It is through songs such as "I Know an Old Lady" that many children first learn about the food chain: The spider swallows the fly, the bird eats the spider, the cat eats the bird, and so on. (It is also the first exposure of many children to a more adultlike vocabulary word, "absurd" [placed to rhyme with "bird" in the second verse].)

Ubiquitous also in every culture are the kinds of knowledge songs that encode information vital to the survival of every member of the group, not warnings about crocodile aggression, but day-to-day guides such as how to cook certain dark green leaves so that they are less bitter, or where to get fresh drinking water without invading the territory of a neighboring tribe:

Over here is where we get our water
Over here is where we get our water
Where the large-winged birds drink
Once a long time ago the father of Erdu
Went to the watering place over there
Where the women of the Baklata go
And the men of the Baklata killed him
We never go there, we never go there
Over here is where we get our water

Early in our history, songs like this would have encoded knowledge about which foods were safe to eat and which weren't, probably in the sort of singsong rhymes we know today as the yellow school bus songs. These knowledge songs were essentially "how to" songs: how to skin an animal; how to make a spear, a water jug, or a watertight boat. We see versions of them today in pop music, in songs such as "How to Save a Life" by the Fray, "The Locomotion" by Little Eva (and later by Grand Funk Railroad), "The Heist" by Busta Rhymes ("that's how we make movies"), or the whimsical "How to Build a Time Machine" by Aussie singer Darren Hayes.

Some of these knowledge songs would naturally but unintentionally have encoded things that weren't true, of course—superstitions or folk theories. Superstitions are nothing more than inaccurate conclusions drawn from observations, experience, or hearsay. They are mentioned in contemporary popular songs as diverse as Devo's "Whip It" ("step on a crack/break your mother's back"), Janet Jackson's "Black Cat," Keith Urban's "A Little Luck of Our Own" ("Black cat sittin' on a ladder/Broken mirror on the wall"), and the Rolling Stones' "Dandelion" (with the implicit message that to blow on this particular plant will give you the power to foretell the future).

With the invention of the printing press, the need for knowledge songs started to fade. In preliterate societies, they were the sole repository of cultural knowledge, history, and day-to-day procedures. They would have been fundamental to information transmission. Today knowledge songs are of a different stripe. The most well-known today is the alphabet song—every child in Western culture learns it. ("Thirty days has September" has musical aspects of rhyme, even though it is usually chanted rather than sung.) But new knowledge songs are being composed all the time. The children's television show of the 1990s *Animaniacs* featured songs that a generation of kids used to learn such things as the states of the United States and their capitals (set to the melody of "Turkey in the Straw") and the nations of the world (set in rhyme to the tune of the "Mexican Hat Dance"). The impressive thing about the latter composition is that composer Randy Rogel not only got the countries to rhyme, but mentions 160 of them more or less according to geographic region. Although he doesn't mention *every* nation in the world, he excludes only a relatively small number of lesser-knowns, and even manages to slip in a joke:

United States, Canada, Mexico, Panama, Haiti, Jamaica, Peru
Republic Dominican, Cuba, Caribbean, Greenland, El Salvador too
Puerto Rico, Colombia, Ven-e-zu-e-la, Honduras, Guyana and still
Guatemala, Bolivia, then Argentina, and Ecuador, Chile, Brazil . . .

Norway, and Sweden, and Iceland, and Finland, and Germany, now in one piece

Switzerland, Austria, Czechoslovakia, Italy, Turkey, and Greece
Poland, Romania, Scotland, Albania, Ireland, Russia, Oman
Bulgaria, Saudi Arabia, Hungary, Cyprus, Iraq, and Iran

Ethiopia, Guinea-Bissau, Madagascar, Rwanda, Mahore, and Cayman
Hong Kong, Abu Dhabi, Qatar, Yugoslavia . . .
Crete, Mauritania
then Transylvania
Monaco, Liechtenstein
Malta, and Palestine
Fiji, Australia, Sudan

Several of my undergraduates every year gleefully sing to me the "Parts of the Brain" as they learned them from this show, set to the melody of "De Camptown Races," with a high-pitched voice chiming in the words "brain stem" where the "doo-dahs" would go: "neocortex, frontal lobe (brain stem, brain stem)/hippocampus, neural node, right hemisphere." By combining laboratory experiments with real-life ethnographic studies, psychologists are just beginning to understand how these and other parts of the brain are able to encode and preserve so much information in music.

In the 1930s, Albert Lord and Milman Parry recorded folk songs in the mountains of (then) Yugoslavia (but not Crete, Mauritania, then Transylvania). For hundreds of years or longer, traveling singers there, mostly Muslim, have gone from town to town, staying for a few days at a time and singing oral histories that can be thousands of lines long and take several evenings to complete. Experienced singers in this tradition might know thirty to a hundred such epics. Some of them memorize their songs with very high accuracy: In one case, a singer heard a song once and then brought it into his repertoire; tape recordings seventeen years later show astonishing consistency, with only very minor errors of wording.

The Gola of West Africa place a particularly high value on the preservation and transmission of tribal history. Part of the reason for maintaining the histories is practical. The knowledge of kinship origins can help establish among contemporaries familial connections and the reciprocal responsibilities attendant to those; being able to claim a relative when you have no food can mean the difference between life and death in a subsistence culture. In addition, ancestors are viewed as distinct personalities who continue to exert influence over the course of present events, even when they've been long dead. Much of this rich oral history is held in musical form by oral Golan historians.

The ancient Hebrews set the entire Torah—the first five books of the Old Testament—to melody and recalled it from memory for more than a thousand years before they ever wrote it down. Even today many Orthodox rabbis can sing every word by heart. The so-called Oral Torah, the commentaries, instructions, emendations, and explanations contained in the written Talmud, have also been memorized verbatim by many, also set to music.

Songwriters know implicitly that setting something to music is the best guarantee that it will be remembered. In contemporary society, writing things down on paper, a PDA, or a computer may seem more practical, but it may not be more powerful. Songs stick in our heads, play back in our dreams, pop into our consciousness at unexpected times. Oliver Sacks tells the story of a song he could not get *unstuck* from his head, Mahler's *Kindertotenlieder* (songs on the death of children). Unable to identify the piece—which grew to create in him a sense of "melancholy horror"—Oliver sang it for a friend. "Have you abandoned some of your young patients, or destroyed some of your literary children?" the friend asked. "Both," Oliver answered. "Yesterday I resigned from the children's unit at the hospital . . . and I burned a book of essays." Oliver's mind had brought up Mahler's song of mourning for the death of children, a reflexive, inventive, and subversive way of symbolizing the previous day's events.

When Genesis sings "All this time, I still remember everything you said," or when Bryan Adams sings "I remember the smell of your skin/I remember everything/I remember all of your moves," they are encoding their personal memories in song. By bonding the sentiments to evocative chords, rhythms, and melodies, they are tapping into a millenia-old and powerful way to save primitive, pure emotional reactions to feelings and events that are important to them. The music itself, even without words, can be very effective at evoking the right sentiment. (Research shows that people are very accurate at identifying the intended emotion of a piece of novel instrumental music.) Adding in the words and their interaction with melody, harmony, and rhythm can print the message indelibly in memory for a lifetime; even longer as new generations of listeners hear the song, learn it, and pass it on.

Some knowledge songs are written as a plea for the listener to remember something specific and important to the writer, as in "The Last Song," by indie rockers Sleater-Kinney:

I need you out of me before I turn into you
I can't stand to look at you,
Until you remember everything

The recipient's only hope for saving the relationship is to "remember everything," and the song serves as a reminder that this is what (s)he needs to do.

Often songwriters are not asking for others to remember, but write *for themselves* to remember. Take one of Johnny Cash's most famous songs, "I Walk the Line":

I find it very, very easy to be true
I find myself alone when each day is through
Yes, I'll admit that I'm a fool for you
Because you're mine, I walk the line

On the surface, it seems like a sweet love song, sung by the man to his sweetheart back home. But in the third verse, Cash explicitly invokes memory as he pledges to think of her:

As sure as night is dark and day is light
I keep you on my mind both day and night
And happiness I've known proves that it's right
Because you're mine, I walk the line

My reading of this verse is that Cash is not in fact singing the song to *her*, but to *himself*. The ironic underpinning is that in fact he *doesn't* find it very, very easy to be true. He *wants* to be, but it is a struggle. The song's function is to remind *him* why he's doing this—because of the "happiness [he's] known" with her and that he doesn't want to risk. "I Walk the Line" is the song of a man in conflict, on the road, with a wandering eye, hoping against all odds and knowledge of his own weakness that he will be true. "I Walk the Line" is a knowledge song, a song of feelings encoded in song so that he will remember "when each day is through" what he promised himself he would do.

Popular songs such as "I Walk the Line" or Genesis's "In Too Deep" are only a dozen or two dozen lines long. How are

Homerian epics, or the long oral histories and ballads of the Yugoslavians, the Gola, or the ancient Hebrews, remembered? Psychologists Wanda Wallace and David Rubin, among others, believe that the mutually reinforcing, multiple constraints of songs are crucially what keeps oral traditions stable over time. In most cases, it turns out, the songs are not remembered verbatim, word for word. Rather, broad outlines of the story are remembered, perhaps using visual imagery, and structural constraints of the song are memorized. This is a much more efficient use of memory than pure rote memorization of the words, using up far fewer mental resources. In Chapter 1 I spoke about the importance of form in poetry, and in song, form is *the* critical feature that helps to recall lyrics.

The mutually reinforcing, multiple constraints that help us to remember song lyrics are principally rhyme, rhythm, accent structure, melody, and clichés, along with various poetic devices such as those we saw in Chapter 1, including alliteration and metaphor.

The rhyming scheme we find in most songs constrains the words that can appear in the last position of rhyming lines. Even though there may exist several words rhyming with the correct word, semantic constraints will prevent most of those words from working in the context of the song. Say, for example, I know a song that begins:

The sun goes down, old friends drink and chat
The wind's in the trees, the dog growls at the ——.

Although words such as "hat," "fat," and "mat" could fill in the blank and maintain the rhyme, our brains reject them almost instantly, unconsciously, because they are considerably less *likely* than an alternative ("cat") in terms of semantics, not to mention in terms of our life experience of storytelling norms in our culture, in which cats are far more frequently mentioned in the same sentence as dogs than are hats, chats, mats, gnats, wine-making vats, or baseball bats. All work syntactically and poetically; one is far more likely semantically.

In Rodney Crowell's song "Shame On the Moon," suppose you just can't remember the word at the end of the first line in the second verse, and all you can remember is that the lines rhyme with each other (a structural memory):

Once inside a woman's heart, a man must keep his ——
Heaven opens up the door, where angels fear to tread

There are perhaps a dozen or so words that rhyme with tread: bed, bread, red, shed, but none of these seem likely in terms of semantics. Even if you can't remember the correct word ("head"), your brain can figure it out more or less on the spot during reconstruction of the song. In fact, there now exists a wealth of research that what we think of as our ability to remember is grossly overinflated, and that many dozens or hundreds of times a day we are *creating* bits and pieces of recollections on the spot. Our brains stitch these creations together seamlessly with our actual recollections and we're none the wiser. In the jargon of cognitive science, this is called the constructive aspect of memory, and it happens so often, so spontaneously, and so quickly that we usually don't know which parts of a recollection are faithful copies of our memory and which are simply plausible inferences our brains made for us. If bits and pieces of a song lyric can be generated on the spot, according to rules, then less of the song actually has to be remembered, and this is far more efficient for the brain.

We engage in this kind of parsimonious and rule-based memory retrieval all the time in other domains, for example in memorizing phone numbers. I have several friends in San Francisco and I've memorized their seven-digit phone numbers. When I go to call them, I remember that they're in San Francisco, and that the area code for San Francisco is 415. Instead of having to memorize an extra three digits for all of my friends, I memorized the *rule* (San Francisco: add 415 to the number). I didn't do this consciously, this isn't some sort of mnemonic *strategy* I employed—it is automatic.

An example of a constructed memory is if I were to ask you about the last time you went to a restaurant. You might respond that you walked in the door, were met by a host or hostess, shown to your seat, where you were given a menu, your order was taken, your food was brought, and so on. Now suppose your narrative doesn't mention the server bringing you the bill at the end of the meal, and I ask you, "Did the server bring you the bill after you were finished eating?" In laboratory experiments like this, many people say yes, they remember being brought the bill, but there exists ample evidence that they aren't actually *remembering,* they're merely *assuming* (or "constructing" the memory in the jargon of cognitive science), because this is part of our shared, common knowledge about what types of things usually happen in restaurants. We don't have to remember all the events for any specific restaurant visit, because some of them are so similar and tightly scripted (and indeed, we tend to remember only if something distinctive occurs, such as the server bringing you the *wrong* bill).

Interestingly, we engage in a very similar process when we recall a conversation, piece of text, or speech, using what psychologists call *gist* memory. We tend to recall the gist, but rarely the precise words. Again, this is a demonstration of parsimony in memory and the fact that in a world of constantly changing environments, literal recall is seldom important. Most messages then are encoded using only a few words or concepts, and our knowledge of the English language and how to form sentences allows us to recreate something *similar* to what was said—close in meaning, but perhaps different in specific expression. We also do something analagous with melodies. When I was in a high school marching band, we played John Philip Sousa's famous march, "The Stars and Stripes Forever." There are several points in the piece where the wind instruments play a low note, followed by a rapid flurry of notes all leading up to a much higher note. It would be inefficient and also unnecessary to try to memorize every single note of that flurry. Instead, research shows that in cases like this, musicians typically memorize the low note, the high note, and how many beats there are available to get from one to the other. Then, using their knowledge of scales and tonality— *rules*—they construct the intermediate notes as and when they're needed.

The very poor typical recall of text stands in stark contrast to the very good typical recall of song lyrics, especially in the case of long epic ballads and the information-set-to-music that I call knowledge songs. Again, this is because songs provide form and structure that jointly serve to fix and constrain possible alternatives. We don't need to store every word of the lyric in our brain's memory banks; we only need to store *some* of them, along with the story and knowledge of the structure of the song. Structural knowledge may include things like the rhyming pattern (the fact that lines one, two, and three all rhyme while four doesn't; or the fact that line one rhymes with line three and line two rhymes with line four, and so on).

If all this seems far-fetched, this is only because it happens unconsciously and automatically, not subject to introspection. Most brain processes, when reduced to neuroanatomical or cognitive description, seem far-fetched, and this is because evolution has created a number of illusions with respect to thought, illusions that serve adaptive purposes. The most elaborate and largest illusion evolution has given us concerns consciousness itself, something I'll return to in Chapter 7. It may also seem, based on introspection, that your brain is *not* generating all the possible rhymes for a forgotten word, but research has shown that this is in fact what's happening, and that in the first five hundred milliseconds or so, dozens of alternatives are considered by your brain (subconsciously) and then filtered through a sieve of cognitive constraints.

Songs also have a rhythm, of course, and this constrains the syllables that can be comfortably squeezed into a given amount of time, and thus again limits the possible words when we don't recall each and every one of them. Take the first line of "I've Been Workin' on the Railroad." If you forgot the name of the song and whatever that thing is you've been workin' on all the live-long day, and the lyric brought you to the dead end of "I've been workin' on the blank-blank," it is relatively clear from the rhythm that a two syllable word is what's missing. If you sing "I've been working on the tra-acks," it sounds funny because the two-note melody there doesn't really support an elongated one-syllable word. A phrase longer than two syllables, such as "the Union and Pacific Rail Line," seems too crowded.

The rhythm interacts with melody and accent structure during the word "railroad." The higher note sounds like it is accented by virtue of its position in the rhythm, falling on a strong beat (the "one" in a four-beat measure of music) while the lower note falls on a weaker beat (the "three"). In this way the rhythmic structure implies a word whose accent is on the first syllable, like "ráil-road," as opposed to say, "gui-tár," whose accent is on the second syllable.

Clichés, or less pejoratively, common word combinations, also help us to remember lyrics. Expressions such as "I'll love you until the end of time" or "letting the cat out of the bag" are so common that if we hear (or recall) only a few of the words, the rest of them follow. For example, you might remember only the first four words of this lyrical phrase: "We used to fight like cats and ——." Even a child could fill in the missing lyric, because this phrase occurs so often in ordinary speech. In fact, this exact phrase "fight like cats and *dogs*" is found in dozens of pop lyrics including songs by Dolly Parton ("Fight and Scratch"), Paul McCartney ("Ballroom Dancing"), Harry Chapin ("Stranger with the Melodies"), Tom Waits, Nanci Griffith, and Indigo Girls ("Please Call Me, Baby"), and Phil Vassar ("Joe and Rosalita"). Another common idiom, the phrase "spill the beans," shows up in songs by six artists that couldn't be more diverse: Yes ("Hold On"), Lard ("Pineapple Face"), DJ Jazzy Jeff & the Fresh Prince ("I'm All That"), Boomtown Rats ("Tonight"), Carly Simon ("We You Dearest Friends"), and Smash Mouth ("Padrino"). In all these cases, songwriters are taking advantage of the brain's ability to retrieve sequences of pre-stored information when cued with only a small piece.

Poetic features of lyrics, such as assonance or alliteration, also help to bring up words from memory; if we remember that the song has these features, we again don't necessarily have to store every word. Great feats of memory are all about parsimony, cognitive economy—using rules that will generate dozens or hundreds of correct answers on the spot rather than memorizing and having to recollect each item.

Some songwriters flout these customary principles, and this itself can become a memory aid. When Paul McCartney sings "Hey Jude/Don't make it bad/Take a sad song . . . ," each word falls right on a melody note in perfect time, just as you would expect. But on the final line of that first verse, he makes a "mistake," one that sounds odd, singing: ". . . and make it bet-ter-er-er," stretching the second syllable of the word "better" out over four notes. On first listening, it is jarring. But we remember it for its distinctiveness. Even if you forget the word "better" (or fail to encode it, although neither possibility is likely, because the very distinctiveness of this compositional move virtually guarantees it will become solidly encoded in memory), you can re-create the word just by remembering that there was something funny going on there, a two-syllable word stretched out to four syllables. Given the semantic constraints of the text before, there just aren't that many words that can fit in that final slot. (Paul uses the same technique later in the song, of course, stretching out the word "be-gi-in" to three syllables.)

The individual effects of rhyme, rhythm, accent structure, melody, cliché, and poetic device can be subtle. The idea of mutually reinforcing constraints is that the effects are additive—no single effect is always enough to help us generate a missing word, but together they can turn an incomplete memory into a near perfect performance. The interaction of these different cues allows the lyrics of ballads and other knowledge songs to remain relatively stable over centuries.

In Wallace and Rubin's studies of expert ballad singers, they found that even when the singers made word errors, they seldom misremembered the structure of the song—the rules describing its invariant features. The end-rhyme sound, the number of beats per line, and the number of lines per verse remain relatively constant within a particular ballad, and the singers studied rarely made errors about these. Ballads in general are stabilized by the common characteristics they share—well-known, stylistic norms within a given tradition whether it is Yugoslavian, Golan, Indonesian, North Carolinian, or Ancient Greek. These common characteristics are so embedded in the idiom that when singers of a given tradition are asked to write a *new* ballad, encoding a new event, they tend to employ all of the same tools and incorporate the same structural characteristics of that form.

In one song Wallace and Rubin studied extensively, "The Wreck of the Old 97," the *form* of the ballad was clearly serving as a memory aid. "Words and music," they write, "are intertwined. The words have a metrical pattern, which must correspond to the rhythmical pattern, the beat structure, and the time signature of the music. . . . In this ballad, the meter consists of two unstressed syllables followed by a stressed syllable. Usually the unstressed syllables are shorter in length than the stressed syllable. . . . The meter and rhythm are closely related in that the number of stressed syllables equals the number of beats in the music. . . . Thus, the music and words constrain each other."

If these multiple and reinforcing structural constraints are such an important ingredient in remembering song lyrics, it follows that a song with fewer of them would be recalled with more errors. Wallace and Rubin created just that situation for an ingenious experiment—they changed twenty-four words in "The Wreck" to eliminate assonance, alliteration, and rhyme. Notice that they specifically made changes that affected *poetic* characteristics of the words, the weakest of the structural constraints I listed above. (No end rhymes were altered, and the number of syllables in each word and the stress patterns were maintained.) The unaltered and the new version of the song were then taught to people who had never heard it before. Analyzing the words that had been changed, Wallace and Rubin found that singers were more than twice as likely to recall a word verbatim if it possessed these poetic elements (the original version) than if it did not (the altered version). This shows just how constraining—or *helpful*—the relatively weak factor of poetics is.

As of this writing, a popular television show in the United States features hapless contestants misremembering song lyrics, which on the surface may seem to undermine my arguments. First, what makes the show funny is that millions of *viewers* remember the lyrics that the contestants forgot and can't understand how they could have forgotten them. Second, understand that network television is entertainment, and that the producers undoubtedly go out of their way to pre-select people with a bad lyrical memory; they may represent only a minority of the population, but they are the ones that make for the most entertaining spectacle. Third, recall that the average person today is exposed to tens of thousands of songs. The average oral historian may have only been responsible for memorizing and retelling thirty or forty in the days before radio. Finally, our ability to recognize that we've heard songs and to repeat portions of them should not be confused with the potential ability we all may have to truly commit songs to memory when we focus our attention and energy on doing so; asking people to sing back songs that they may have only ever heard in the background at a shopping mall or on a car radio is not the most rigorous test of their abilities (and not a scientific experiment!).

Now here comes the amazing part of Wallace and Rubin's study (and why they are my heroes). It's based on the principle of constructive memory—that we don't actually remember all the details we think we do, we fill in many of them subconsciously by making plausible inferences. Wallace and Rubin looked at the errors made by those singers who were asked to learn the altered version of the song (and heard only that version). Many of them spontaneously, and without prior knowledge, *recovered* the poetic words that had been removed by the experimenters. In other words, the stylistic norms of assonance and alliteration—the tendency for them to be prominent in these sorts of epics and ballads—was something that the subjects in their experiment had encoded in their long-term memory. When they tried to remember the song—involving a process of *reconstructing the words*—they "remembered" the words that originally belonged in the song. For example, if the original song contained the alliterative phrase "real rough road," and the altered version was rewritten and learned as "real tough road," some singers (wrongly?) sang it back as "real rough road."

In family and tribal histories, ballads commemorating important battles, and so on, minor errors in wording are probably so insignificant as to be not worth mentioning. There are at least two cases, however, where errors are completely unacceptable. One is those knowledge songs that were created to preserve precise, practical information about how to do something, such as make a watertight raft, prepare food or medicine that might otherwise be poisonous, or accomplish some other action where both the order of the steps and the details of the procedure are critical. A second case is those knowledge songs that are intended to convey religious information, where every single word must be preserved verbatim because it is considered sacred and divinely given—there are strong cultural pressures to recall such material accurately or not at all. In cases like these, humans exhibit a remarkable ability for virtually perfect recall. How could this be?

When people accurately recall texts of great length, they are typically texts set to music—songs. It is far rarer for such prodigious memory to be demonstrated with straight, musicless text, and the reason for that appears to be the theory of multiple, reinforcing constraints. Insight into the matter comes from another very clever experiment by Rubin.

Rubin asked fifty people to recall the words of the Preamble to the United States Constitution (try it yourself now before reading any further; the words are in an endnote at the back of the book). This of course does not have music, but that's the point—the kinds of errors that people made here were profoundly different than those made by people trying to recollect songs. One group of people remembered the first three words ("We the people") and then . . . *stopped.* You hardly *ever* see this in song recollection. When they forget the words, people singing a song typically continue on humming or singing la-la-la or something else, along with the melody that continues, uninterrupted, to play in their heads, and they pick up words here and there along the way, sometimes recalling entire stretches of the song at a later point down the line. Another group of Rubin's subjects recalled the first seven words ("We the people of the United States") and then quit. Nearly every one of his fifty subjects stopped cold when they failed to recall *the next word* in the sequence, whatever that word was.

The places in the text where people got stuck were far from arbitrary: 94 percent tended to get stuck at phrase boundaries, natural breaks where one would pause to take a breath. Rubin replicated these findings with the Gettysburg Address. Songs, because of their rhythmic momentum, are far more easy to "keep on going" in our heads without the words; this gives our brains a chance to jump in whenever a word or word fragment becomes available again, making lyric recall in songs typically better than lyric recall without them. This is one reason why the average person probably has a more intimate and emotional connection to music than to poetry—because he or she can recall more of it, and more effortlessly.

It's important to look in more detail at *how* rhythm plays its role in aiding recollection of lyrics. When we recall a song, rhythm provides an internally consistent hierarchy of temporal units—syllables form poetic feet, which in turn form lines, which form phrases, verses, or stanzas—and these rhythmic units usually coincide with the units of meaning in oral traditions. These rhythmic units with their accent structure create positions in the music where words must go—they don't permit us to omit *part* of a rhythmic unit, thus conspiring to preserve the integrity of lines and verses. (If any omission is to occur, it would generally be at the level of a large-scale repeating unit such as a verse, and then different mnemonic processes come into play to maintain *its* integrity.)

Recall from Chapter 1 that a mark of most effective poetry is that it has its own rhythm, a music about it that we hear when we recite it out loud. If poetic rhythm, combined with the sorts of other poetic devices we looked at in Chapter 1, really helps us to remember poetic text, you might expect that people would have better memory for poems than straight text, and moreover, that their errors would be more similar to those we find in music recollection. Both of these turn out to be true. David Rubin asked undergraduates to recite from memory the 23rd Psalm, which contains many poetic elements but (in the English version) lacks rhyme. Here, most undergraduates started up again at a later point in the psalm after they stopped (and it was usually at the beginning of a new section that they came in). It seems as though the internal rhythms of the psalm kept on playing in the students' heads, and they jumped in whenever they could recall a word. (Don't forget that the Psalms, as originally composed, were set to music.)

Additional evidence for the notion that people "play back" such recollections in their heads comes from my own experiments with Princeton computer scientist Perry Cook. We found that when college undergraduates sang their favorite songs from memory, they tended to sing them at almost exactly the right tempo. If they forgot words, they kept going and jumped in later, again at what *would have been* the right place if they'd kept on going, as though the band kept playing and they simply came in at the next appropriate moment. And from their own subjective reports, all of them had vivid mental imagery of the music. They weren't so much trying to reproduce the song from memory as *singing along* with a track in their heads.

Getting back to the memorization of text, when very long word sequences are involved, we typically resort to two tried-and-true mnemonic techniques: rote memorization and chunking. Rote memorization is simply reciting a sequence back over and over again (often in the quiet privacy of our own minds) until we've got it. This is how most of us learned our multiplication tables in grammar school, the Pledge of Allegiance, the Gettysburg Address, or the Preamble to the Constitution. The interesting thing here, though, is that not all words are created equal in rote memorization. Some words take on more importance because of the expressive emphasis we are taught to give them, some because they evoke particularly pleasant or vivid imagery, and some because they contain certain (preplanned, on the part of their writers) poetic qualities. Internal rhymes, assonance, and alliteration of a Cole Porterish quality help to reinforce the Gettysburg Address for example (maybe because assonance makes the heart grow fonder):

Four *score and seven years ago*
Our fathers *brought* forth

We see the repetition of the f sound here acting as a mnemonic, as well as the repetition of the long o sound in *four, score, ago, forth*.

When the text we're trying to memorize is more than a dozen or so words long, we tend very naturally and without coaching to break it up into bite-sized, more readily memorized, sanitized and organized units, or chunks, and then stitch these chunks together later. This is also how musicians memorize pieces they perform. With the exception of special neurological cases of people with photographic memory (or the auditory equivalent, what I call *phonographic* memory), most musicians, dancers, actors, and other performance artists, young and old, do not sit down and learn a new piece from front to back all at once. They concentrate on getting a small part of it just right and then they learn another small part. They then spend some time learning the transitions from one part to another. The evidence for this process remains long after the piece has been committed to memory and its performance is flawless: Actors who have to redo a take often ask to go back to the beginning of the line, paragraph, or scene. Musicians return to the beginning of the phrase. Bring out the score, point to an arbitrary note where you'd like them to start in a memorized piece, and most musicians will ask to start somewhere else, at the beginning of some chunk that they learned. When musicians make errors in prepared pieces, these errors provide additional clues to the way the piece was originally learned. It is more common for a musician to skip an entire section (a failure to remember how the chunks were stitched together) than for a musician to skip a note or short group of notes within a section. *Between*-section errors are far more common than *within*-section errors. Not all notes or words are equally salient.

The Gettysburg Address is substantially easier to memorize than multiplication tables, which typically require pure rote memorization. Personally, I remember having great difficulty with these in grammar school. I made a little card with them written down, and while I walked to school or had a spare minute, I would glance at the card and test my memory. With no rhythm or melody to attach the numbers to, pure brute force of repetition was required: two times two is four; two times three is six; two times four is eight. I got up to the sixes without too much difficulty, but well into high school I couldn't remember the "twelves" unless I started from a part of the tables that had personal meaning for me, the one that represented my height in inches at the time I was originally trying to memorize them: Twelve times four is forty-eight. If I wanted to recall twelve times eight, I'd have to start my "poem" at this salient point and work my way up: twelve times four is forty-eight; twelve times five is sixty; twelve times six is seventy-two; twelve times seven is eighty-four; twelve times eight is ninety-six. Because of incessant teasing from my neighbor in third grade, Billy Latham (an excellent drummer at the time, by the way), who had learned *his* tables all the way up to twelve times twelve, I had also memorized the one that Billy taunted and drilled me with most frequently: twelve times nine.

The reality of these chunks has been demonstrated many times in psychology laboratories. Asked to sing song lyrics from memory, beginning from an arbitrary point, people have great difficulty. Asked even to answer simple questions about lyrics they know, people are influenced by the hierarchical structure of the lyrics—revealing in the laboratory certain organizational properties of human memory. Here's an example. Does the word *my* appear anywhere in the lyrics to the song "Hotel California"? What about the word *welcome*? Both words do appear, *my* is the ninth word in the song, and *welcome* is the ninety-sixth word. But people take *longer* to say yes to the first question than to the second, and psychologists believe this is because *welcome* is the first word of the chorus, a privileged position in the hierarchy of your memory for the piece.

Most North American children learn the alphabet by learning the letters set to the melody of "Twinkle Twinkle Little Star" (the same melody as the beginning of "Ba Ba Black Sheep"). The song has phrase boundaries because of its rhythmic structure, gaps between the letters *g* and *h*, *k* and *l*, *p* and *q*, *s* and *t*, and *v* and *w*, forming natural "chunks":

abcd efg hijk lmnop qrs tuv wxyz

Indeed, most children don't memorize this all at one sitting, but rather they work their way up, memorizing these small units. The rhyming scheme helps too: The ends of all the chunks rhyme with each other except chunks three and five. Even though as adults we know the letters of the alphabet (or think we do), many of us still rely on that song when searching for the specific location of a letter. In one experiment, it was found that it takes college undergraduates much longer to say what letter comes just before *h*, *l*, *q*, or *w* than before *g*, *k*, *p*, and *v*. Crossing the chunked boundary carries with it some cognitive cost.

Notwithstanding what I said about errors at phrase boundaries, some professional musicians and Shakespearean actors do indeed have perfect recall for a memorized string and can begin anywhere; this ability probably develops from *overlearning* processes, and under stress it may disappear. Something like it shows up in even semiprofessional or amateur actors and musicians when they need to start from the very beginning of a piece. Effectively, their "stitching together" of subcomponents has worked so well that the entire piece, though at one time consisting of small units, has become a single memory trace, not easily disassembled, occasionally indestructible, lasting a lifetime, and sometimes outlasting even the memory of their own family members' names.

The use of chunking and the occasional inviolability of long, overlearned sequences is not confined to contemporary Western society. The Greeks discussed these ideas in formulating instructions for mnemonics two thousand years ago, and the anthropologist Bruce Kapferer (from the University of Bergen in Norway) has observed them in researching the myths of Sri Lanka. In trying to form a catalog of different demons and their characteristics in a cultural group he was studying, he would ask a local oral historian to describe the myth of a particular demon. The historian would respond that the exact details were contained in a certain song. "I will sing it and you tell me when the demon you want has his name mentioned. Then I will go slow so that you can put it onto your tape recorder." The myth information was stored in the song, and the song was known only sequentially, and only from the beginning.

Getting back to the Greeks, the 2,500-year-old *Iliad* and the *Odyssey* represent great feats of memorization, without music, but with clear poetic, rhythmic constraints doing much of the work and thus creating less demand on the brain. Their prosody is very tightly constrained. As just one example, the number of syllables per line is almost always constant, and the last five syllables in a line are almost always long-short-short followed by long-short. The ordering of short and long syllables, and the preferred locations for word breaks are formulaic—not just any word will fit the rules. For example, words containing a long-short-long or short-short-short syllabic structure can't be used in Homeric epic at all. Obviously, if one has learned the rules for such a form, opportunities for inserting the wrong words are extremely limited.

According to Jewish tradition, the complete Torah (the first five books of the Old Testament) was completely memorized by Moses, and then taught to the elders and leaders of the Hebrew people in the Sinai Desert, who in turn taught it to the million or so people who left Egypt as part of the Exodus, sometime around 1500 B.C.E. We know that the Hebrews had written language (the tablets of the Ten Commandments were written), but on Moses's strict instruction, not one word of the Torah was to be written down, and for more than one thousand years the history, knowledge, religious customs, and practices were reportedly handed down only through oral transmission. And the form of that oral transmission, according to all accounts, was song.

Jewish mystics believe that the very *sound* of the words will bring divine favor, even if the speaker doesn't understand the words themselves. Similarly, in the Zoroastrian tradition, it is believed that the soul (Urvaan) can be reached by the specific vibrations that come from chanting the Avesta Manthras. It's not just the meaning of the prayers, but also their *sound* that matters, for the "attunement" of the soul. In Zoroastrianism, Staota Yasna is the theory of auditory vibration. That's why the prayers are still recited in the original language, Avesta, even though it isn't spoken any more.

The Qur'an was also set to rhythm and melody—a chant—and learned in a similar fashion, although it is explicitly not considered "music" and is structurally very different from Arab music we hear today; in fact *singing* the Qur'an is strictly forbidden. The Qur'an itself describes the means of its recitation (*tarteel*) in verse four: "and recite the Qur'an in slow measured rhythmic tones." The power of song to aid memory is evidenced in a following fatwa against singing. Islamic scholars believe that "music and singing carrying obscene content, instigating to sin, lechery, destroying noble intentions and leading into temptation, are inadmissible (*haram*). . . . The degree of inadmissibility becomes higher if an obscene vocabulary acquires a musical accompaniment that contributes to a better remembering and thus enhances its impact."

In the case of the Torah, the melody itself contained clues not only to the words, form, and structure of the narrative, but also to interpretation of words or passages that might otherwise be ambiguous. That is, the assignment of words to melody (and vice versa) was not arbitrary—it helped not only as an aid to memorization and recall but also to ensure the correct interpretation.

This sort of interplay is not at all unusual 3,500 years later. In the song "Superstar" as sung by the Carpenters (written by Leon Russell), Karen Carpenter sings the line "Long ago, and oh so far away" using a vocal technique that artfully reinforces the meaning of the words. She delays the pronunciation of the word "far," reinforcing the idea of distance. While holding the word "away," she brings out a subtone in her voice that conveys the sense of deep loss and separation. In Steve Earle's "Valentine's Day," the singer is surprised by the arrival of the day, and realizes too late that he has forgotten to get his girlfriend a present. He writes her a song instead. The appearance of surprising and nonstandard chords underscores the meaning of the words, adding tension and deeper meaning to the lyric.

The reasons for the insistence that the Torah be transmitted orally are a matter of speculation. One proposal is that the success of the Jewish people, one of the oldest continuously living civilizations in the world, is due to especially close ties between parents and children and the bond created through the oral transmission of knowledge—knowledge which, in the case of the Torah, intimately binds family history, moral lessons, political history, codes of daily conduct, and instructions for maintaining an orderly, just society. If the information had been written down and learned by reading, the knowledge transmission would have flowed in one direction, from book to student. The oral transmission enabled—virtually *required*—interaction, questioning, participation; what the physicist-turned-Torah-scholar Aryeh Kaplan called a *living teaching*. Indeed, the ancient Hebrew scholars wrote that "the Torah is meant to be alive, to be spoken." Like the poetry we encountered in Chapter 1, it is meant to be *heard*, both in the ears of the people and in the minds of those who have learned it and can play the song of it back in their heads at will, in times of scholarship, trouble, or praise. It's also been proposed (somewhat less logically) that the restriction on writing the Torah down existed because some knowledge of customs and traditions would be lost if it was committed to writing—that the sum total of the knowledge known by the people exceeded what could be written down. When the rabbis decided sometime between 150 B.C.E. and 200 C.E to commit all the teachings to writing, much debate and disagreement indeed ensued, about many details. (All of the debate is captured in the Talmud—in fact, that is primarily what the Talmud *is*—a record of what were essentially judicial proceedings and deliberations about what precisely the oral teachings were and how they were to be interpreted.)

From a memory standpoint, the cantillation (as the Torah melody is known) provides the same sorts of constraints that other songs do—perhaps even more—facilitating the memorization and preservation of an enormous amount of text. Without recordings, however, it is impossible to know for sure how well the original words of sacred texts such as the Torah and the Qur'an were preserved through their oral transmission. We don't know, and *can't* know, the extent to which melodies changed, rhythms were rewritten, emphases were altered. But the fact that different subgroups of contemporary Jews sing different melodies suggests that there was no one magic formula for preserving the information—time and tide would have caused minor changes as in the child's game of telephone, and over generations these differences could have become considerable, and they would have become amplified. Humans are a highly adaptive species. As we moved to new locations, to communities with their own musical cultures and traditions, original melodies may well have become altered or distorted by the influence of local songs. Even the prosody of the new languages (the "music" of the language) has been shown to influence the songs of that linguistic culture. As the Mongols entering southern Europe on horseback, the Armenians dispersing to Paris, and Italian-Americans crooning in Hoboken found, local sounds pull at the immigrants' long-preserved melodies and rhythms, creating new hybrids that continue the cultural evolution of their songs, at the possible expense of losing some of their original (melodic and textual) information.

The existence of different melodies for Torah today suggests that errors may indeed have crept into the text when it was transmitted orally—if melodies can change, so can words. (In fact, much of the discussion during the compilation of the Talmud in the first few centuries C.E. acknowledged that some errors by then had *al-*

ready crept in, and concerned how to resolve those errors.) Indeed, the discovery of the Dead Sea Scrolls reveals that multiple versions of the sacred texts exist. From both a cognitive and a theological perspective, the errors are mostly relatively minor and unimportant, of the type we saw in Rubin's study of ballads.

Across all these examples, a common thread emerges: Knowledge songs tell stories, recount an ordeal, a saga, a particularly noteworthy hunt—something to immortalize. The demonstrated power of song-as-memory-aid has been known to humans for thousands and *thousands* of years. We write songs to remind ourselves of things (as in Johnny Cash's "I Walk the Line") or to remind others of things (as in Jim Croce's "You Don't Mess Around with Jim" or Fleetwood Mac's "Don't Stop"). We write songs to teach our young, as in alphabet songs and counting songs. We write them to encode lessons that we've learned and don't want to forget, often using metaphor or devices to raise the message up to the level at which art meets science (rather than simple observation), making it at once more memorable and more inspiring, as Andy Partridge does in "Dear Madam Barnum" (performed by his band XTC). The name Madam Barnum is clearly made up, meant to portray her as the ringleader in some abusive emotional circus she has subjected the poor songwriter to, and from which he now hopes to extricate himself:

I put on a fake smile
And start the evening show
The public is laughing
I guess by now they know
So climb from your high horse

And pull this freak show down
Dear Madam Barnum
I resign as clown

Songwriters often invoke a well-known tale or legend in the context of a new song. In "Dear John," the writer of this song (Aubry Gass, and as sung by Hank Williams) is again encoding an experience into a song, presumably so that he won't forget that it was his own misbehavior that caused his woman to leave him; he weaves into the message two well-known Old Testament references:

Well when I woke up this mornin',
There was a note upon my door,
Said "don't make me no coffee babe,
'cause I won't be back no more,"
And that's all she wrote, Dear John,
I've sent your saddle home.

Now Jonah got along in the belly of the whale,
Daniel in the lion's den,
But I know a guy that didn't try to get along,
And he won't get a chance again,
And that's all she wrote, Dear John,
I've fetched your saddle home.

Note the interesting shift to the third person in the second verse ("I know a guy that didn't try to get along"), a conceit to make us think that it isn't actually he who was left by a woman, a move that underscores the point that he's sending the message out to others as a warning: Don't do what I did; treat your woman right.

Hard-won lessons are a staple of knowledge songs, from Paul Simon's "Run That Body Down" to Ani DiFranco's aptly titled "Minerva" (after the Roman goddess of knowledge) to the Magnetic Fields' "You Love to Fail." As with Aubry Gass's, the songs seem to be simultaneously directed from the songwriter to him- or herself and to all of us. Guy Clark, one of my all-time favorite songwriters, brings a lifetime of lessons seemingly learned the hard way into his song "Too Much," a romp that is made all the more fun (and memorable) by the form he imposed on himself: Every line of the verse begins with the same two words ("too much"), compiling a litany of everyday pleasures, too much of which will cause the various calamities specified at the end of each line:

Too much workin'll make your back ache
Too much trouble'll bring you a heartbreak
Too much gravy'll make you fat
Too much rain'll ruin your hat
Too much coffee'll race your heart tick
Too much road'll make you homesick
Too much money'll make you lazy
Too much whiskey'll drive you crazy
. . .
Too much limo'll stretch your budget
Too much diet'll make you fudge it
. . .
Too much chip'll bruise your shoulder
Too much birthday'll make you older

Part of what makes the song memorable is the obvious *fun* that the composer had in writing it, reflected in the joy the performers bring to playing it. The sense of whimsy is enhanced by decomposing the familiar idiom "carrying a chip on your shoulder" to yield "too much chip'll bruise your shoulder." In "too much limo'll stretch your budget," he taps into our memory associations of limos and "stretch limos" to use both senses of the word *stretch*. (It is no wonder that Clark is a favorite of many of the best songwriters in the business, including Rodney Crowell, whom Guy mentored.)

A song like "Too Much" turns the memory process into a game—the first part of each line cuing the second part or vice versa. If we forget the line, logic can often deliver it to us just as it does in "I Know an Old Lady Who Swallowed a Fly." It is worth making the cultural point that these sorts of songs are found in every society we know of. And the fact that children tend to love these songs is evidence that our ancestors found this sort of mental play both rewarding and an efficient form of learning and information transmission.

Up until this point, I've been considering songs as they are recalled and sung by one individual at a time. But knowledge songs—from Huron's yellow school bus songs to Torah cantillation—are more typically sung by groups of people. In this context their position as a foundation of culture and their durability become even more apparent. I've already described the social bonding that comes from synchronous music making, and the neurochemical effects of singing, but there are manifest cognitive benefits that are conferred to the group-as-a-whole, apart from any benefits to the individual when people sing together. Group singing shows a special ability in retrieving information that a lone individual might not be able to recall, an *emergent property*. Emergent behavior occurs when groups can do things that individuals cannot. Ant and bee communities are examples of emergence where intelligence arises out of a multiplicity of relatively simple and seemingly unmotivated actions. No single ant "knows" that the hill needs to relocate, for example, but the actions of tens of thousands of ants result in the hill being moved, efficiently, effectively, even "intelligently." The Stanford biologist Deborah Gordon writes, "The basic mystery about ant colonies is that there is no management." No ant stands at the periphery of the colony directing traffic.: "Hey you! Stop playing 'rub-the-feelers' with that worker ant and get a move-on! Come on guys, break it up, there's enough moldy peanut for every— HEY BUDDY! What're you, taking a break? Go help those guys carrying the heavy praying mantis carcass!" With no ant in charge, how on earth do ants get anything done?

Ant colonies exhibit behavior very similar to that of other systems with a very large number of units or components, all of which interact, and the consequences of whose interactions change over time. Physicists call these *nonlinear dynamical systems.* (They're called *nonlinear* because the effects of these interactions can't simply be added up, they sometimes have to be expressed as powers, or other higher mathematical functions. They're called *dynamical* because the influence of one event at the beginning can have profound effects later in time as that initial effect is carried forward.) In systems like these—which include rain forests, stellar transits, the stock market, and even the faddish propagation of hit songs—small, seemingly chaotic, and unrelated behaviors can end up having large effects as they interact, spread, and develop over time. In other words, extraodinarily simple individual units—like ants, neurons, atoms, or musical notes—can generate complicated and often counterintuitive global behavior.

On the surface, you might think that groups remember knowledge songs better because they spread out the memory burden among a greater number of people. But it's not the case that ten singers each remember a different line—there is no prior agreement, no coordination of the learning or recall. Instead, due to a variety of differences among individuals (and these differences can be genetic, environmental, based on IQ, personal motivation, personal taste, and random factors) some people are going to remember some parts of a song better than they remember other parts. There may be nothing systematic about it, and the remembered parts could even change from day to day and week to week.

But something special happens when a group starts to sing together—something extraordinary from a cognitive (and dynamic complex systems) perspective, something you've probably experienced yourself in any place where people come together to sing: football games, church, campfires, or political rallies. On your own, you might get stuck after the first line of the song. Singing with a single friend, your companion might remember the first word of the second line and that keeps you going for another few words, but then neither of you can remember the third line just now. In a large group, *no one* has to be able to recollect the entire song. Just one person singing the first syllable of a word can trigger a recollection in another group member to bring the second syllable of that same word, which in turn can cue a group of people to that whole word and the next three words after it. Imagine this notion propagated through a large group of dozens or hundreds of people, and throughout every syllable of the song—a sort of group consciousness emerges in which no single member of the group can be said to know the song, but the group itself does.

Even in a non-emergent system, if someone *misrecollects* and offers a wrong syllable, musical note, or word, he will likely be overpowered by the greater number of people who correctly remember it. (This is a version of Oliver Selfridge's famous connectionist *pandemonium model* of human perception.) The reason for this is that, although it is relatively likely that any part of the song will be misremembered at any given moment by any member of the group, it is relatively *unlikely* that all of the misrememberers will misremember it in the same way. The correct rememberers will tend to outnumber any of the various schools of misrememberers (and with a large group, it becomes increasingly unlikely that a substantial number of individuals will forget at exactly the same time and in the same way). In a *dynamical* system, every second new information is revealed and this influences the future development of the system. At least some of those who misremembered will correctly remember when given the right cue. This is the fundamental mechanism by which extraordinarily long stretches of textual information have been preserved, handed down, and communicated across hundreds or thousands of years. Errors do of course creep in, but the more tightly constrained the poetic/musical form of the song (such as we see in Homerian epics or even in the twelve-bar blues), the more likely the message will arrive intact, unaltered, and resistant to future distortions.

Nonlinear dynamical systems are thus characterized by (1) local propagation of information among individuals (a neighbor providing the right lyric) through a (2) nonlinear mechanism (individual and group memory and cognition in this case) involving (3) individual variability (e.g., heterogeneous song recollection). These three properties lead to (and are required for) the emergence of low and asynchronous occurrence of errors in the whole group despite the high probability of error observed in single individuals.

The existence of *group* memory and *group* singing may itself have been selected for by evolution, which may have favored those individuals who could bond into groups for the purpose of collective action. I believe that this *group selection process* and the survival and reproductive advantages conferred upon members of large groups (as opposed to lone individuals) is fundamental to how human societies were eventually formed. While synchronized singing positively affects the psychological state of individuals, synchronous occurrence of errors in singing has to be avoided at all costs. An individual must balance self-confidence in her singing compared to a willingness to align with her neighbors. This trade-off is itself nonlinear and dynamic, changing throughout the course of a performance (and it is found in many other dynamic systems, such as ecosystems).

The power of knowledge songs to encode and preserve information, to engage even young children in their recollection and transmission, points to an ancient evolutionary basis for them. Viewed from a larger context, knowledge songs can be seen as a special instance of art, and in particular, of the kind of art that seeks to inform. Art and science are seen by many as inhabiting opposite ends of a continuum, a line that runs from abstraction to specificity, or from romance to logic. I've spent my life in pursuit of knowledge in both domains, and surrounded by people who have pursued either or both. Many musicians I've known pursue their music in the kind of systematic, deliberate, *studious* fashion that could only be described as scientific: Frank Zappa, Sting, Michael Brook, and David Byrne for example. Others pursue it more intuitively, including Carlos Santana, Jerry Garcia, Billy Pierce, and Neil Young. This is not to say that the latter four haven't *worked* at it, but their approach to the work strikes me as based more on feeling than on any sort of system. Bill Evans, my favorite pianist, sums up the latter approach:

"Words are the children of reason and, therefore, can't explain it [music]. They really can't translate feeling because they're not part of it. That's why it bugs me when people try to analyze jazz as an intellectual theorem. It's not. It's feeling."

I've come to see art and science as occupying two ends of a continuum that wraps around on itself like a circle, so that the two meet at a common point. Both art and science involve perspective-taking, representation, and rearrangement—the three foundational ingredients of the musical brain. When these three are combined, we get metaphor (letting one object or concept stand for another) and abstraction (letting a hierarchically larger concept stand for subordinate elements). Both art and science rely on metaphor and abstraction because they take sensory, sensual, and perceptual observations and distill them to an essence. In both, we can get more meaning out of a single piece of information than if that information were delivered in its raw, literal form. Art and science are about extracting and abstracting world knowledge in a form that makes it more readily understandable and memorable—what they share is a sense of overview and unifying themes, decisions about which facts-of-the-world are relevant and which are not. Art and science are not able to represent *everything*; instead they entail (require) difficult choices about what is the most important.

Science is a not the simple reporting of facts—that's only the preliminary step in any scientific investigation. Real science, the kind that offers a parsimonious and predictive understanding of how the world works, involves taking those facts and generalizing global principles from them; abstraction is required for this, as is creativity, rationality, intuition, and a sensitivity to form, similar to what is required in the creation of long-lasting art. It may be self-evident that music requires these things, less so perhaps that you can't have science without the musical brain.

A single painting of a sunset tells us how the sunset felt to the artist and can convey those feelings forever. A mathematical model of the movement of the solar system (including the material constitution of the sun, and local environmental conditions) allows us to predict whether a given sunset will be spectacular, run-of-the-mill, or completely obscured by clouds. Both can inform our behavior and serve our memories, both grab us at the nexus of feeling and thought, emotion and interpretation, brain and heart.

Knowledge is emotion. Some people say that science just *is,* that it is merely a collection of facts and measurements that exist outside of the realm of emotion and caring. But I disagree. Of the millions (perhaps infinity) of possible facts about the world we memorize, document, and pass on to others, we *select* those that we think are important, and this is an emotional judgment. We are motivated to care about some and not others, and as we saw earlier, emotion and motivation are two sides of the same neurochemical coin. It is true that the mere fact that $2+2=4$, or that hydrogen is the lightest element we know of, is without emotional content. But the fact that we *know* these things, have gone to the trouble to learn them, reflects interests, priorities, motivation—in short, reflects emotion. Scientists are motivated by intense curiosity and a desire to interpret and represent reality in terms of higher truths—to take collections of observations and formulate them into a coherent whole that we call a theory. Of course artists do the same thing, taking *their* observations and trying to formulate *them* into a coherent whole that we call the painting, the symphony, the song, the sculpture, the ballet, and so on. Knowledge songs are perhaps the crowning triumph of art, science, culture, and mind, encoding important life lessons in an artistic form that is ideally adapted to the structure and function of the human brain. We need to know. And we need to sing about it.

Science, like nature,
Must also be tamed

With a view towards its preservation.
Given the same
State of integrity,
It will surely serve us well.

Art as expression,
Not as market campaigns
Will still capture our imaginations.
Given the same
State of integrity,
It will surely help us along.

The most endangered species,
The honest man,
Will still survive annihilation.
Forming a world
State of integrity,
Sensitive, open and strong.

(From "Natural Science" by Rush)

I've been dropping the new science and kicking the new knowledge
An M.C. to a degree that you can't get in college
. . .
It's the sound of science

(From "Sounds of Science" by the Beastie Boys)

Thanks for the ride. Big Science. Hallelujah.
Big Science. Yodelay-hee-hoo.

(From "Big Science" by Laurie Anderson)

CHAPTER 6

Religion

or "People Get Ready"

When I was four years old, my grandfather took me to Kearny Street in the heart of San Francisco's Chinatown. My "cousins" Ping and Mae—the technicians who developed the X-rays in his radiology office—met us there. As he usually did, Ping lifted me up onto his shoulders and walked me around as I held onto his forehead, my hands sometimes excitedly slipping over his eyes and blocking his view. And what there was to see! Dancers in pink and purple costumes ran up and down the street, firecrackers exploded, floats decked out in fresh and plastic flowers carried waving local dignitaries, and traditional Chinese music came bursting out of loudspeakers, bullhorns, instruments, and mouths everywhere we went. The whole crowd was smiling, laughing, jumping, celebrating as free-spirited, barely coordinated parts of a single organism. I had never seen so many happy people in one place at one time. And it was infectious. When Ping lowered me to the ground, Grandpa and I danced in place, and he swung me around in circles, my feet lifted off the ground by the swirling centrifugal force. Mae gave me a whistle to blow and a pin to wear on my T-shirt. At home and at synagogue, we sang songs on Friday nights and we had even sung something a couple of months earlier at Jewish New Year, but those songs were solemn affairs, slow and tedious, nothing like the Chinese songs. Ceremonies didn't have to be somber!

Human rituals around the world have many elements in common, suggesting either a common origin to them all or a common biological heritage. Some are joyful and some serious, some disciplined and others performed with structured abandon. When we break down these activities into their elements, we see a remarkable continuity with activities in the animal kingdom, strongly suggesting that evolution had a hand in guiding us toward the particular ways in which we express ourselves through movement and sound. The common conception that humans possess abilities that make us uniquely human—language is often trotted out as a crowning achievement, with religion and music not far behind—is sharply contradicted by some of the newest research in neurobiology. Animals are indeed capable of many of the things that only ten years ago we thought were our species' sole inheritance, the abilities falling along a continuum rather than appearing abruptly in *Homo sapiens*. What is different is our species' ability to discuss and plan these activities with self-conscious awareness of them, and to bind them in time and location to particular beliefs. Animals may perform rituals, even quite elaborate ones, but only humans commemorate and celebrate, and only humans tie these to a belief system. When the Edwin Hawkins Singers sing "Oh Happy Day," they celebrate the day Jesus "washed away sin" with some of the most joyful and uplifting emotion ever recorded. No animals celebrate a particular date, a birth, or commemorate a decisive battle—to do so requires brain structures that they may possess but do not use the way we do.

The continuity of behaviors from animals to us bears scrutiny. Ants and elephants bury their dead. Humans mourn theirs, and typically with elaborate ritual—sometimes solemnly, sometimes joyfully, almost always accompanied by music. Neanderthals were burying their dead long before *Homo sapiens* walked the earth, but the archaeological record suggests that their burials were an accidentally adopted behavior for hygienic reasons—like cats covering their feces; no traces of ornaments, jewelry, or other accoutrements accompany Neanderthal burial sites, whereas these are almost always present in human graves. Humans built on this existing physical action of burial to imbue it with a cultural and spiritual component. Ceremony, as a uniquely human invention, commemorates important events. These can be events of our human life cycle such as birth, marriage, and death, or events in our environmental life cycle such as the seasons, the rains, daybreak, and nightfall. Rituals tie us to the event itself, and to the cycle of history in which many similar events have previously occurred and will continue to occur. They are a form of externalized, social memory, and when marked by music, they become even more firmly instantiated in both our personal and collective memory. The songs, sung at the same time and place every year (in the case of seasonal or holiday songs), or at gatherings commemorating similar events (funerals, weddings, births), bind these events together in a common theme, in a common set of beliefs about the nature of life. The music acts as a powerful retrieval cue for these memories precisely because it is associated with these and only these times and places.

The evolutionary changes that furnished humans with the musical brain—an enlarged prefrontal cortex, and all the myriad bilateral connections with cortical and subcortical areas—formed a crucial step in social development of our species. With these evolutionary changes came self-consciousness (an aspect of perspective taking), which brought with it spiritual yearnings and the ability to consider that there might be things more important than one's own life. I believe a particular kind of music—songs associated with religion, ritual, and belief—served a necessary function in creating early human social systems and societies. Music helped to infuse ritual practices with meaning, to make them memorable, and to share them with our friends, family, and living groups, facilitating a social order. This yearning for meaning lies at the foundation of what makes us human.

Like music, religion is found in all human societies (and for both, people disagree as to whether they have an evolutionary or a supernatural basis). In spite of great differences in beliefs and practices and in geographical location, no known human culture lacks religion. This strongly suggests that religion is more than a meme—information transmitted to people through culture—and may have an evolutionary basis. Émile Durkheim, one of the fathers of sociology, taught us a century ago that anything that is universal to human culture is likely to contribute to human survival. Modern biologists extend this notion to animal behavior, looking for links across species as a way of understanding the evolution of the brain. Behaviors that we consider quintessentially human don't just fall out of the sky from nowhere, but are distributed along a continuum of behaviors that are remarkably similar to those we see in animals (and so presumably contributed to animal survival). It may not be possible to cleanly distinguish ritual from religion, and perhaps the distinction is not as important as understanding the continuous nature of how they relate to one another, of how rituals became bound up into religion in the first place.

Rituals involve repetitive movement. Many animals show ritualized behavior, such as dogs circling a few times before lying down, birds swaying from one leg to another, or raccoons washing their faces after a meal. What separates these from human rituals is a cognitive component, human self-consciousness: We are aware (most of the time) of our behaviors, and we assign them a higher purpose and sense of meaning. We wash our hands because of *germs*. We light candles to mark an event. We talk about our rituals and sing about them. What sets religious practices apart is that they are *sets* of rituals, bound to a common narrative or worldview. That is, rituals exist in a part-whole relationship to religious practice.

The anthropologist Roy Rappaport defined ritual as "acts of display through which one or more participants transmit information concerning their physiological, psychological, or sociological states either to themselves or to one or more of their [fellow] participants."

The *display* aspect is critical—rituals serve as a form of *communication*. Also critical is the inclusive nature of his definition, which explicitly allows for the display to be self-reflexive, to serve only the person engaged in the ritual. Thus, an individual washing her hands prior to preparing ceremonial food, or laying sticks at a fire altar, or even musicians executing a set of scales as a pre-concert warm-up—all signify to the doer that she is becoming ready (through this preparatory ritual) to advance to the next stage of a plan or operation.

Rappaport defined religion as "sets of sacred beliefs held in common by groups of people and . . . the more or less standard actions (rituals) that are undertaken with respect to these beliefs." He defines *sacred* as those beliefs that are unverifiable through normal physical means or through the normal five senses—the belief or faith in things that are not corporeal but that can influence the course of our lives.

Religious ceremonies and practices almost always incorporate ritual behavior—repetitive motor actions: bowing seven times, making the sign of the cross, folding and unfolding your hands in a particular way. Anthropologists have identified certain features of human religious practices, believed to be universal, applying across disparate cultures, times, and places:

1. Actions are divorced from their usual goals. We may wash parts of the body that are already clean, talk to others who are not evidently there, pass a piece of fruit from hand to hand in a circle (when clearly the goal is not to pass the fruit *to* someone, but just to engage in the act of passing), walk around a stone exactly four times, or perform actions that do not have an immediate tangible goal.
2. The activity is typically undertaken in order to get something: more rain, more yams at harvest time, to heal a sick child, to appease the gods.
3. The practices are typically considered compulsory. Community members consider it unsafe or unwise (or improper) not to perform them.
4. Often no explanation is given about the form of the activity. That is, while all participants may understand the goal of the ritual (e.g., to influence the gods), typically no one provides an explanation of how *these particular actions* will yield the desired outcome.
5. Participants engage in behaviors with more order, regularity, and uniformity than in their normal lives: They line up instead of walking or standing anywhere they please; they dance instead of moving; they greet one another with special signs,

gestures, or words; they wear similar or special clothing or makeup.

6. Objects are taken from the environment and infused with special meaning, sometimes by piling, ordering, stacking, or aligning them.
7. The environment is restructured and delimited—a holy circle, a taboo area, a special place that only the elders or the pure can enter.
8. There is a strong emotional drive to perform the activity, and anxiety is experienced if it is not performed (or if participants feel it hasn't been done properly); individuals feel a sense of relief when it is completed.
9. Actions, gestures, or words are repeated—perhaps three to ten times or more. The exact number is crucial to proper observance, and if the wrong number is used, the performer starts over.
10. There is a strong emotional drive to perform the ritual in a particular way—the actions are somewhat rigidly interpreted and defined. Someone in the community—perhaps an elder—is known to perform each activity best, and others try to emulate that example.
11. The rituals almost always involve music or rhythmic, pitch-intoned chanting.

These features are found in the religious ceremonies of Muslims, Hindus, Christians, Jews, Sikhs, Taoists, Buddhists, and Native Americans, as well as hundreds of ceremonies of preliterate and preindustrialized societies. And the story of ritual is intimately bound up with music—which almost always accompanies it—and with human nature. Whether you believe that man invented religion or received it from God is not the issue, and I don't want to get distracted by that question here. The remarkable similarity of human religious rites to one another, and to certain animal rituals, can be invoked as evidence for either view. Several recent scientific studies have shown that there exist neural regions that might be called "God centers": When they are electrically stimulated, people report having intense feelings of spirituality and communication with God. Some scientists argued overconfidently from these findings that religious belief must be "merely" a product of the brain, and that therefore humans must have invented God.

I told all this to my good friend, Hayyim Kassorla, a learned and respected Orthodox rabbi, and without missing a beat he snapped back, "*So what* if there's a center in the brain that makes people think of God? Why wouldn't there be? Maybe God put it there to help us to understand and communicate with him." My mother, an observant Jew, added, "The similarity of human religious practice across cultures is because God found that these practices work and so he gave them—with some variation—to all peoples." The point is that we can consider the biological, evolutionary, and neural evidence for ritual behaviors without necessarily impinging on anyone's personal beliefs about the origin of the universe or spirituality. We don't need to resolve the physical/metaphysical question before proceeding with the evolutionary one.

Ritual behavior is evidently innate and hardwired in humans. Most children enter a stage of development around age two, peaking around eight, in which they show phases of ritual behaviors: perfectionism, collecting, attachment to favorite objects, repetition of actions, and even a preoccupation with the ordering of things—a stage of "it has to be done this way" in which children may line up their toys or organize their environment in particular ways. Young girls hold tea parties for their real and imaginary friends. The table is set *just so*. The guests have to sit in assigned places. The hostess can become agitated when things are not organized or consistent with her own internal notion of the proper ritual order: "You sit here, he sits here. No! Mr. Rabbit gets his tea *first*!"

Spontaneously—without explicit instruction and without ever having heard of it from someone else—many children connect their ad hoc rituals to the supernatural or to magic, and they imagine effects the rituals might have on a variety of outcomes, from the weather to telekinesis to getting their way.

Of course I didn't engage in tea party rituals as a child since I was a boy—my ritual phase was more automotive and involved seat belts and the seat belt song. In 1961, when I was three, the Ad Council of America launched a program of public service messages on television about the importance of "buckling up for safety" with car seat belts. A catchy jingle exhorted us to tell our parents that there was a correct order, a sequence of events that must be respected: *Before* driving, buckle up your safety belt. I remember hearing that jingle and singing it all around the house. Seat belts were new in 1961 and most cars didn't have them; my parents had learned to drive without them. Although our family car, a Simca, was equipped with them, my parents hadn't gotten used to using them, and probably weren't convinced of their effectiveness. (There hadn't been any crash test dummy experiments done yet.) My mother tells me that whenever she and my father got in the car, I would sing the little jingle and get furious if they drove even a few feet without having buckled up. I was deep in my "correct order of things" phase.

In children, rituals tend to be associated with anxiety states—fear of strangers, the unknown, attack by strangers or animals, and possible contamination. This leads to bedtime rituals such as checking for monsters, wanting to hear bedtime stories, or holding your special fuzzy blue blanket. The ritual adds a sense of order, constancy, and familiarity that psychologists believe counteracts the uncertainty and fear of the unknown dangers. Oxytocin—the trust-inducing hormone that is released during orgasm and communal singing—has been found to be connected to the performance of ritual, suggesting a neurochemical basis for why rituals have a comforting effect.

These behaviors are so widespread, and occur so regularly in childhood, that they no doubt have an evolutionary and genetic origin. A drive toward creating symmetry, toward lining up and ordering one's environment, is present even in birds and some mammals. From an adaptation perspective, this order makes any intrusion by an outsider immediately and clearly visible. Those of our ancestors who found pleasure in hand washing or in creating symmetrical protective borders around their encampments may have been more successful at fending off both micro- and macroscopic threats to their health and safety, and passed on a desire to do so to us through the oxytocin system. Richard Dawkins's observation is compelling: No one of us alive today had an ancestor who died in infancy. Every one of our ancestors lived long enough to pass on his or her genes to us. While we can't say that every minute behavior of our ancestors was adaptive, none of them could have been grossly maladaptive, to the point of causing early death, or of making them fatally unattractive to a member of the opposite sex. The ubiquity of ritual behaviors does suggest they served, in some form, an important survival function.

Some ritual behaviors become uncontrollable, and when that happens nowadays we diagnose it as obsessive-compulsive disorder (OCD). Some researchers have speculated that impaired dopamine and GABA regulation—and the consequent abnormal control of the brain's "habit circuits" in the basal ganglia—lead to OCD in both humans and animals. The basal ganglia store chunks or summaries of motor behavior, and when they are inadequately regulated, one only finds emotional satisfaction when they are allowed to run in a loop of repeated activity, over and over again.

Rituals in animals and adult humans tend to be born from many of the same concerns as children's rituals—purity, contamination, safety, but also mating. To claim that human rituals, even when encased within sacred religious systems, are *uniquely* human, is to ignore the rich repertoire of animal rituals that resemble them. As just one of many possible examples, consider the mating ritual of the Australian bowerbird (of the family *Ptilonorhynchidae*). While it may strike us as elaborate and complex, it is in fact no more so than the mating rituals of hundreds of other species of birds, mammals, amphibians, and fish. Once a year, the male birds spend several days gathering brightly colored objects such as feathers, shells, and berries to create bowers, elaborately decorated structures, usually shaped like a pathway, a hut, or a small pole. After the bowers are completed, the males sing and dance, concluding a successful ritual by selecting a female to mate with (and the female typically chooses the male based on the quality of his bower and his singing and dancing).

Compare this with an annual ritual of the villagers on Pentecost Island in the South Pacific nation of Vanuatu. Each year young males take part in the ritual of *nagahol,* as part of a plea to the gods to ensure, among other things, a good yam crop. Young males construct tall, colorful, elaborately decorated poles up to seventy-five feet high and as all the villagers sing and dance, the males climb the poles and then jump off, held by a thin vine. If he concludes his dive successfully, a male is then considered an adult and can take a wife from among the spectators. As far as we know, there are no bowerbirds on Pentecost Island and never have been. Either a common neurobiological imperative underlies the ceremony, or it is a coincidence. Is *nagahol* part of a religion or is it an isolated ritual? At what point does ritual become religion? Sulawesi villagers engage in a rain dance that includes stylized imitation of the sound of rain, meant to induce it. Their dance and music and actions have a clear goal and intended effect. I leave the distinction between religion and ritual in the mind of the beholder (and the question of whether this belief constitutes a *belief system* or not is inessential to a discussion of the evolution of religious song).

Religion—whether God-given, man-made, or a gift of natural selection—can be seen as an important part of inclusive fitness. All higher animals have a "security-motivation" system that monitors environmental conditions and motivates them to act via emotional states if danger is imminent. The system monitors both external events in the world and internal states such as pain, fever, and nausea. The brain mechanisms underlying this can be broken down into three parts: (1) an appraisal system compares events being monitored to a list of things known to be dangerous (either known by experience, or innately); (2) an evaluation system determines the magnitude of the hazard; and (3) an action system causes the human or animal to execute a response that will reduce the danger, by moving, running, fighting, or any number of other strategies innate or learned.

The display aspect of single rituals, or those sets of rituals that become bound into religious acts, allows common human fears and concerns to be given a broad social context in which they can be shared with the community and make better sense.

Religion further lets us partition our fears into those that we and our community are going to worry about and those that we are not, and to take collective action toward addressing the former and systemically, in a formally sanctioned way, ignoring the latter. Depending on our belief system, we might decide as a community to pray for the health of a loved one, but not for dead relatives to come back to life; modern Christian rituals focus on Jesus not on Zeus or Thor, and we have society's permission to ignore the latter two gods and any demands we fear they may have.

When we pray for the health of a dying loved one, the termination of the prayer confers a great psychological advantage: It allows us to stop worrying. We breathe a sigh of relief and affirm that "it's in God's hands; her fate is decided." This is clearly adaptive, because it propels us to get on with our lives and to stop ruminating about things that we can't change, and to worry about those we can. Interestingly, however, the fear-security-motivation system was "built" thousands or tens of thousands of years ago and it is not responsive to those current threats that are most dangerous: This is why so many of us are afraid of spiders and snakes, which cause far fewer deaths than cars or cigarettes, which most of us are not afraid of.

Another function of rituals is to change the state-of-the-world, thereby reducing ambiguity. Consider male puberty rites, which are part of most cultures. Unlike the onset of menses, which unambiguously signals the transition from girlhood to womanhood in females, no such biological marker exists for males. Male puberty rites remove the ambiguity about the young male's role in society, whether he should act as a boy or a man. The rite turns more-or-less information into yes/no information: Before the rite he is a boy, after he is a man.

A marriage ceremony turns a man and a woman into husband and wife. This parallels a prominent theory in psycholinguistics about human speech acts. Most of the time our utterances are simply speech acts that express opinions, make requests, provide information, or share our emotional state. There is a small class of utterances, however, that hold a special status as being able to change the state-of-the-world. This happens when a duly recognized official makes a declaration that has either legal or definitional consequences. An example is a minister saying "I now pronounce you husband and wife." If he is officially recognized by the church or state, this simple sentence changes the status of the couple. Similar state-changing utterances include a judge announcing a verdict (the guilty or innocent proclamation dramatically changes the legal and practical state of the defendant), a government official deputizing a law enforcement officer, the chief justice inaugurating the President, or even a coroner pronouncing someone dead. (Note that in the latter case, even if the person isn't dead at all, the proclamation by a duly authorized coroner changes the legal status of the victim, allowing for autopsy, burial, and other actions not otherwise allowed.)

When people come together and want to inaugurate or celebrate a change of social status, such as a wedding or the installation of a leader, music is virtually always there. Music is also there at harvest celebrations and anniversaries of a birth, death, or important battle. Commemoration seems to require music. The specificity of time and place is an interesting aspect of religion songs (and the subset of ritual songs I'm including) that sets them apart from the other five categories in this book. Everyone knows ritual songs can only be sung at the right time and place. But songs of joy, for example, can be sung anytime that songs can be sung; not in a library or in the middle of a play, for example, but if songs are otherwise permitted, there is no reason that you couldn't sing a joy song, or a friendship song, or a knowledge song. Religion/ritual songs, on the other hand, and their associated religious/ritual events, carry very strict restrictions on time and place.

Take for example, the Elgar composition "Pomp and Circumstance," also known as "The Graduation March," played at high school and college graduation ceremonies all across North America, as students file up to receive their diplomas (it is also New Zealand's national anthem). The song has interesting musical qualities. It begins with a legato line of notes in close proximity to one another: The first fifteen notes are all stepwise and then the sixteenth note of the piece takes a large ascending leap of a perfect fourth immediately followed by a falling perfect fifth—a move that grabs our attention. "Pomp and Circumstance" is played with a stately pace, and the instrumentation gives it a sense of majesty, seriousness, and procession. So well known and uniquely associated with commencement, it is even used at some nursery school and kindergarten graduation ceremonies. But no one plays this piece at sporting events, or on a dinner date, or at a wedding.

Even the most insensitive clod would recognize its misuse at the wrong time or place. If there was a high school assembly being conducted, an informational session perhaps, for students who were being held back a year in school due to poor academic performance, and the principal played "Pomp and Circumstance," it would seem cruel. Or consider a Ph.D. oral exam. It would be unusual, but not unacceptable, for a student to start playing a tape recording of "Pomp and Circumstance" once the exam was over and the committee told him he had passed. But if the student played the same song *before* the exam, the committee would find it so presumptuous as to be offensive.

The criticality of time and place is a hallmark of ritual songs, and it is so important that if it is violated, jobs can be lost and even—in the extreme—so can lives. Consider songs that accompany a country's ruler, such as "Hail to the Chief" or "God Save the Queen," played respectively when the President of the United States or the British queen enters a room. If a scheming, conniving underling were to instruct the military band to play the song every time *he* walked in a room, it would be seen as a direct and aggressive challenge to the reigning ruler's authority. In a dictatorship, playing the leadership song for the wrong person could easily result in a death sentence. Such is the importance of time and place in ritual and religion songs.

Ritual and religion songs are therefore, to my way of thinking, bound to particular times and events, and for the explicit purpose of accompanying, guiding, or sanctifying a specific spiritual act. Under this definition "Jingle Bells" or "Deck the Halls" are not religion songs, although they spring from the celebration of a religious holiday. Rather, I see them as friendship songs, binding us to friends and family who hold similar beliefs. Christmas carols can be sung during a broad range of occasions surrounding the season of the holiday; as I conceive ritual and religion songs, they are far more restricted. Similarly, national anthems and football fight songs, although ritualistic (being played at the beginning of a competition, for example), are really serving a social bonding function more than a religious or spiritual one. "The Wedding March," "The Funeral March," the Mass, the Song of Atonement, on the other hand, *are* religion songs in that they *must* be performed at a certain time and place and they *cannot* be performed whenever one pleases. To do so would seem improper. I could sing "Jingle Bells" or "Over the River and Through the Woods" in the middle of July. It might seem odd, but it would not seem improper, sacrilegious, or disrespectful.

Some form of music accompanies every behavior that even remotely resembles a religious practice worldwide, from the Pentecost Islanders' male puberty ritual, to ancient Egyptian funeral services, to a contemporary Catholic Mass. In a great many ceremonies, there is a clear goal to perform the act *as a community.* Part of music's role then involves the social bonding function of bringing together members of a community in this moment of making the request (for food, rain, health, etc.), the feeling of "strength in numbers" at appearing before the gods (invoking a feature of friendship songs), and part of music's involvement is because it is effective at encoding the particular formula of a request that has worked in the past (invoking a feature of knowledge songs). But songs used in a religious context, while they have these elements of the friendship and knowledge songs described earlier, are a radically different type of song because of their connection to a belief system, and their being tethered to a particular time and place. Also crucial is music's power to encode the details of the ritual—remember that by definition, rituals involve repetitive movements, and music exerts its power here to encode the proper conduct of the movements, synchronously with the music.

Consider the ancient *Devr* ritual of the Kotas, a group of two thousand people who live in the Nilgiri Hills, a region bordering the South Indian states of Tamilnadu, Kerala, and Karnataka. Although unique in its details, it highlights common themes among belief, ritual, movement, and music that exist across all cultures and times.

Devr begins on the first Monday after the waxing of the first crescent moon of winter. Villagers gather wood, prepare special ceremonial clothing, eat only a vegetarian diet, reduce their alcohol consumption, and walk barefoot. They clean and purify their homes using special plants (including branches from the tak tree, which contains a purple ellipsoidal berry that grows on a thorny stem). Designated individuals create and transfer a series of special fires that are conduits for divinity. The village deities (who are said to be present in these fires) inhabit stick bundles in a back room of a *mundkanon*'s (leader in all village god rituals) house called a *kakuy*. When the bundles are put in the fire, the deities can express themselves to the community.

The beginning of the ceremony, *omayn*, is signaled by the unison blasts of the *kob* (a brass instrument) along with flutes and drums. The word *omayn* means "sounding as one" and is similar, of course, to the Jewish and Christian *amen* and the Sanskrit *aum* meaning "it is true" or "we all agree." The gods hear these forceful blasts as an attention-getting invitation to enter the village. Much ceremonial music throughout the world has this attention-getting quality, from the rising perfect fourth of "Pomp and Circumstance" to the sudden appearance of the fifth in Kyrie of the Catholic Mass (on the word *Christe*).

For the next ten to twelve days, Kotas perform instrumental music, dance, and sing to express their joy, unity, and respect for the gods, and to entertain them. Particular songs are used during ritual bathing and food offering, as individuals synchronize their movements to the music. A highlight of the *Devr* celebration occurs when villagers join together to re-thatch the temple roof. As the music plays, they throw sanctified materials onto the roof. To perform the ritual properly, the throwing must be synchronized with the horn blasts from the *kob* players so that the upward motion of the throwing arm is simultaneous with the playing of a piercing tremolo on the highest note on the instrument. Other notes give emphasis to changes in orientation and motion, both horizontal and vertical.

In this and other rituals, music performs a critical, synthetic, and catalytic function. Music synthesizes disparate parts of the motor activities under a single melodic/temporal scheme. It catalyzes the actions by its alternation of tension and release: When rituals are synchronized with music specially designed for the undertaking of the ritual, the music reaches an emotional peak when the activity does, and reaches a resolution and release of harmonic tension as the activity draws to a close. Music guides participants to the proper, rigid, accurate performance of the ritual because motor action sequences can be learned *in synchrony* to the music: During *this* part of the song we raise our arms; during *that* part of the song we fold them.

Children's songs in which participants move parts of their body selectively, and in particular ways, are found in every culture. These constitute practice for coordinating music and movement. In my own childhood, a favorite was "The Hokey Pokey":

> *You put your right foot in*
> *You put your right foot out*
> *You put your right foot in*
> *And you shake it all about*
> *You do the Hokey Pokey and you turn yourself around*
> *That's what it's all about!*

Subsequent verses find us putting in our left foot, our arms, our head, our "whole self," and so on. (In a recent dream, I'm climbing a steep mountain to reach a seer at the top. He emerges from his cave, his long white beard and long hair waving in the breeze. I ask him "What is the meaning of life? What is it all about?" He responds by quoting me the verse above, with a pregnant pause just before the last line, and then beams, "*That's* what it's all about!")

A song many Americans of all faiths learn in Sunday school about Noah and the flood has similar motor synchronization:

The Lord said to Noah, "There's gonna be a floody floody,"
Lord said to Noah, "There's gonna be a floody floody,"
Get those children out of the muddy muddy
Children of the Lord

Chorus:
So rise and shine and give God your glory glory
So rise and shine and give God your glory glory
Rise and shine and give God your glory glory
Children of the Lord

During the chorus, children stand at the word "rise," hold their open palms next to their face during "shine," and shake their palms to emulate a glittering action on the words "glory" ("jazz hands"). I have Muslim and Baptist friends who learned the same patterns. "Itsy-Bitsy Spider" and myriad other hand-eye-sound coordination songs train children to move with music, train us for rituals.

Recent research has confirmed that music is a powerful way of encoding motor action sequences—specific movements that must be done in a particular way. Down syndrome children who are otherwise unable to tie their shoes can learn to do so if the movements are set to song. Military units learn to assemble and disassemble guns, engines, and other tasks using song. Rigidity in the performance of the ritual is enhanced by the music:

Notes and words unfold in a precise sequence and at a precise time, and motor actions are learned to synchronize with them. The music also helps to set the emotional tone, to serve as a memory aid for practice, and to synchronize multiple participants.

Most ritual music has a quality of unison rhythm for this reason, but exceptions exist, the most notable and fascinating being pygmy music, which (to my ear) is a conceptual predecessor of the ebullient, enthusiastic singing in many religious ceremonies, which is perhaps best known in black churches in America. I attended synagogue as a child and even sang in the choir, but as I mentioned above, this was a stern, reserved affair: We always sang in rhythmic unison, and only occasionally strayed into three-part harmony. This was in stark contrast to the Cornerstone Baptist Church Choir (singing "Down By the Riverside") and St. Paul's Disciple Choir ("Jesus Paid It All") that I saw on Sunday morning television. Within such choirs, an ever-changing core group of people sing the nominal melody as others sing, improvise, shout, chant, and rejoin whenever they feel moved to do so. The result is a thrilling and exhilarating musical force that could sow doubt in the most ardently confirmed atheist. In gospel music, as sung in thousands of churches, both the community and the individual are celebrated. The unison and harmony lines of the core melody strengthen the sense of solidarity and community, of shared goals (as stated in the song) and shared history (as evidenced by the singing of a song that everyone knows). The ecstatic interjections, some planned and some spontaneous, affirm the individual as an artistic and meaningful entity, created in God's image—leading to feelings of self-acceptance and self-confidence. As India.Arie sings in "Video," her own merging of hip-hop, funk, gospel, and pop:

I'm not the average girl from your video
and I ain't built like a supermodel
But, I learned to love myself unconditionally
Because I am a queen

When I look in the mirror and the only one there is me
Every freckle on my face is where it's supposed to be
And I know my creator didn't make no mistakes on me
My feet, my thighs, my lips, my eyes; I'm lovin' what I see

In the African pygmy music I'm listening to right now, enthusiastic shouting, wailing, and counterpoint run through the song. Rhythms are kept on shaker sticks and drums, often speeding up and slowing down. For the Mbuti people, the forest is benevolent and powerful, and their music is the language with which they communicate with the spirit of the forest, in order to request food, peace, and health. The aim is to communicate intense joy to the forest, which will return it to them. Good music is seen as the embodiment of social cooperation, as is good hunting and feasting. Bad music embraces laziness, aggressiveness, and disputatiousness and is associated with ill humor, shouting, crying, anger, bad hunting, and death. An ultimate goal of pygmy singing is to oppose the destructive force of death.

Although traces of its asynchronous polyphony are found in modern gospel music, in its pure form it is without peer. Pygmy music is so utterly distinctive as to have earned its own entry in *The New Grove Dictionary of Music and Musicians:*

Its most striking features, apparently common to all groups, are an almost unique wordless yodeling, resulting in dis-

> junct melodies, usually with descending contours; and a varied and densely textured multipart singing. . . . This choral music is built up from continuously varied repetitions of a short basic pattern, which takes shape as different voices enter, often with apparent informality. . . . The frequently clear division of the total cyclic pattern between leader and chorus . . . is absent . . . or obscured by . . . the passing round of what might be regarded as soloistic parts from one to another. Some scholars see in this a reflection of the essentially democratic, non-hierarchical structure of pygmy social units.

In describing the music, rituals, and practices of other cultures as I've done here, my intention is to show the great diversity of religious and ritual customs, and the enormous variety of forms of musical expression. I do not mean to draw attention to practices in a way that is disrespectful to them or to their adherents. Obviously, it is important to remind ourselves that preliterate and preindustrial peoples are not childlike or necessarily less intelligent than we—they live a different lifestyle, hold different beliefs, and have a different education. The pygmies famously resisted efforts by a few unwittingly condescending anthropologists to render them as "primitives." (One pygmy man was tragically captured and put in a circus.) A true story attests to their sophistication and attempts to defend their dignity. When asked by the anthropologist Colin Turnbull to play the oldest song they knew for his tape recorder, a group of rain-forest pygmies sang an impromptu version of "Oh My Darling, Clementine," complete with polyrhythmic drumming, stick-beating, and vocal harmony.

Forms of nonsynchronous singing and chanting exist in much

of the world's religious music, from Sephardic Jewish liturgy, to Muslim, Buddhist, and Hindu chanting. Children typically have difficulty with nonsynchronous music, and rounds are all but impossible for young children, who become distracted by the other parts, until they reach a developmental stage in which they have more volitional control over their own attentional mechanisms (and a more highly developed cingulate gyrus in the frontal cortex), sometime around age six to eight. Complex, nonsynchronous music can thus serve as a marker of intellectual maturity.

In more structured forms of music—especially religious music—a leader sings a line and the choir or congregation echoes it, or answers with a prescribed musical reply. We see this in songs like "Oh Happy Day." In such "call-and-response" music, the response may either be a literal musical and textual repeat (as it is here on the first and second replies) or a melodic variation (as on the third reply):

> LEADER: *Oh happy day!*
> CHOIR: *Oh happy day!*
> LEADER: *Oh happy day!*
> CHOIR: *Oh happy day!*
> LEADER: *When Jesus washed . . .*
> CHOIR: *When Jesus washed . . .*

The folk music and work songs that grew out of the enslavement of African-Americans in the rural south incorporated elements of African music and gospel, and many of them featured a call-and-response form. It was these songs that subsequently formed the bedrock of twentieth-century folk and eventually popular music, where the call-and-response became a staple of sixties and seventies rock, as in "Twist and Shout" by the Isley Brothers:

LEADER: Well shake it up baby
BACKGROUND: (Shake it up baby)
LEADER: Twist and shout
BACKGROUND: (Twist and shout)
LEADER: Well come on baby now
BACKGROUND: (Come on baby)
LEADER: Come on and work it on out
BACKGROUND: (Work it on out)

The call-and-response technique became so famous in pop music that it could even be implied, instrumentally, and evoke the same emotional drama and impact as the literal response. The roots of this are in the jump 'n' jive music of the 1940s, for example in Big Joe Turner's "Flip Flop and Fly," where each vocal line is answered by a saxophone line. In Leon Russell's "Superstar," as performed by the Carpenters (with Richard Carpenter's brilliant arrangement), Karen sings "Long ago" and the orchestral instruments echo her vocal melody, and keep up a vocal call and instrumental response throughout the song. McCartney does the same thing with his piano lines responding to the vocal lines in "Let It Be."

Call-and-response, as a specialized form of nonsynchronous singing, is partly predictable, in that we know *when* the next musical event is going to occur, although we may not know exactly *what* it will be. This balance of predictability and unpredictability gives the performance (as distinct from the underlying composition) a palpable excitement. In less structured forms, such as pygmy music or the religious and spiritual music of many indigenous and preliterate peoples, the unpredictability is increased and along with it the excitement. In musics like this, the rhythmic elements—played on drums, rain sticks, shakers, shells, stones, sticks, and hand claps—typically take on a more regular, hypnotic quality that can induce trance states. Just how music induces trance is not known, but it seems to be related to the relentless rhythmic momentum, coupled with a solid, predictable beat (or *tactus*). When the beat is predictable, neural circuits in the basal ganglia (the *habit* and *motor ritual* circuits), as well as regions of the cerebellum that connect to the basal ganglia, can become entrained by the music, with neurons firing synchronously with the beat. This in turn can cause shifts in brain-wave patterns, easing us into an altered state of consciousness that may resemble the onset of sleep, or the netherworld between sleep and wakefulness, or even a druglike state of heightened concentration coupled with increased relaxation of the muscles and a loss of awareness of time and place. When we're engaged in the music making ourselves, and creating elaborate motor movements, we reach the *flow* state mentioned in Chapter 2, similar to an athlete being "in the zone." When we're not explicitly moving (or merely swaying with the beat), the state is different, more like a state of hypnosis, and differences in brain waves are also observed between the two states.

Like many Americans, I was entirely unfamiliar with these types of music and my own childhood experiences with religious music—in a Reform Jewish synagogue that tried to emulate and assimilate aspects of American suburban Protestantism—exposed me to slow, serious, and joyless music. "Whites are afraid to show their emotions, particularly joy," Joni Mitchell told me. "I think it goes back to the original sin and the biblical accounts of Adam and Eve being *embarrassed*—that has negatively impacted white social interactions for centuries. Most white singers don't have anywhere near the emotion that black singers do—Billie Holiday, Bessie Smith—every single note is invested with all the feeling of human existence. In my younger days, I had my little white girl folk voice, and I didn't know how to put emotion in it, and I also hadn't experienced enough of the world to really express life's emotions fully. Black culture is much more balanced, they put a value on emotion and spirituality. White culture wants to keep all of that stuff quiet and tucked away.

"Some of our deepest feelings come through the spirit," Joni continued. "To the extent that religion is a manifestation of spirit, it really ought to reflect the full range of feelings, especially joy. The ballet I wrote (*Shine*) is Gnostic, because Gnosticism rides the cusp, in a way, of all spiritual thought. It absorbed just about every religion, and put the goddess back in, it was earth-friendly, woman-friendly, and all the things that religion isn't at this time. It saddens me so deeply what we, the Woodstock generation, have done to our planet. And nobody listens! We keep trashing it, ruining it, there won't be anyone left fifty years from now and its purely our Tower of Babel arrogance that has brought us to this. Only humans have the stupidity to destroy their own planet. You talk a lot about 'evolution' in *The World in Six Songs,* but maybe it would be more accurate to talk about humans as the products of *devolution,* of a relentless pursuit of perfection in stupidity and arrogance. Even religion today has lost its ability to pull us out—now it's all warrior gods. My song 'Strong in the Wrong' is a direct attack on these subversions of religion. On the other hand, the Gnostic God is a thing within you whereby you lose your self-consciousness and transcend. It's more like Buddhism, in that way. So the Buddhists in the dance troupe kind of lit up because, it's not like the Buddhists are afraid to do a Catholic dance, but man! the Catholics are sure afraid to do a Buddhist dance."

A highlight of the ballet, which Joni choreographed, produced, and wrote the music for, is a dance to one of her favorite poems, "If" by Rudyard Kipling ("If you can keep your head when all about you/Are losing theirs and blaming it on you . . .")

I always considered Kipling's "If" to be a religious poem, not about what it is to be "a Man" (or a Woman, or an Adult), but how to be more Godlike, more spiritually enlightened. Joni rewrote the text slightly, changing the word "knaves," for example. "And I changed the ending," she explains, "because I wanted the ballet to emphasize wonder and delight; the ability to recharge your innocence is what makes you inherit the Earth. I changed the ending to 'If you could have sixty seconds' worth of wonder and delight'—which are those glimpses of the waking mind, they put you there right in the moment—'then the Earth is yours.' In other words, if you can perceive it; if you can wake up for a minute or a second and seize the damned thing, at that moment you own it. It doesn't matter whose property you're on. You could be walking with the owner of a huge parcel of land, but if you see it and he doesn't, at that minute, who owns it, perceptually, spiritually? There's a lot of meat on the bone of that idea."

Whereas Joni looked back to Kipling for spiritual inspiration, David Byrne mentioned "My Body Is a Cage" by the Montreal-based band Arcade Fire.

My body is a cage
That keeps me from dancing with the one I love
But my mind holds the key

"To me," David says, "it's religious and at the same time anthemic. It gets really big at the end, but it's still very personal. This song's not calling for spiritual or political revolution or, 'we must march and fight' or 'we shall overcome,' or whatever. 'My body is a cage that keeps me from dancing with the one I love, but my mind holds the key.' It's beautiful, but to me it's a little bit backwards; usually it's the other way. Usually it's the mind that's keeping the heart from acting. So it's the mind telling the heart 'No! I'm gonna stop you from indulging in your passions.' And then it goes on: 'I'm living in an age that calls darkness light . . .' It's biblical language, but they apply it a little bit to the personal and the political. It's not one of the social bonding or friendship songs from *The World in Six Songs,* a kind of rousing 'hey, we're all in this together.' It's more like one person's torment, one person's inner experience, which is what makes it such a powerful religion song to me."

"My Body Is a Cage" showcases religion as a struggle not just against immorality—its usual sphere—but against immortality. The conviction that there is something beyond this corporeal existence—a life, a future, beyond what we know and see here. But my body is a cage preventing me from seeing it. My body is a cage preventing my essence from being able to reach out and merge with that of my lover or my creator.

David Byrne spends much of his time listening to music of other cultures, and they have infused and informed his own compositions, as they have for Paul Simon and Michael Brook. A favorite religious song of David's is "Roble" by the Argentinean singing group Los Fabulosos Cadillacs.

Ya cayeron ojas secas
El frio del invierno va a venir

"It has the sort of sweeping melody you find in national anthems," David explains. "The melody begins with a slow build and then just climbs up; it has some really peculiar stops and hesitations where it lets the note just kind of hang over so that you know that there's probably an extra bar or something in there, which has this big emotional crest. And then it comes down. Lyrically, *roble* is Spanish for 'oak,' and the words are basically about how it loses its leaves and then they come back."

Sin resistir, sin dormir
Roble sin fin vos sabes lo que es morir
Solo soñar con la lluvia lo lleva a revivir

The anthemic nature of the melody and the long, slow rhythms transform the lyrics from a literal story about an oak tree to a metaphor—a spiritual lesson about change, growth, perseverance, and renewal. "Without resisting, without sleeping, the oak tree knows what it is to die," David says.

"I can't help but apply it to the Argentine political situation, because these guys are from a generation that grew up in the era where people disappeared for their political views. In Spain, Argentina, Romania, citizens have lived through a period as children, and they remember this repressive situation; that was just the way things were. And then, things open up. I can't help but read that the song is a little bit expressive of that as well."

The *Big Idea* of most religions is that even if things aren't so good now, they'll get better. We see this powerfully rendered in the so-called black spirituals of the South, in "We Shall Overcome," and "People Get Ready (There's a Train A-Coming)" by Curtis Mayfield:

All you need is faith to hear the diesels humming
You don't need no ticket, no you just thank the Lord

Psychologists and anthropologists have found that after a certain minimal level is attained, increased material wealth and comfort do not make people happier. An oft-quoted adage is accurate: The secret of being happy is to be happy with what you have. Too often in Western society—a society built on consumption—we don't stop to enjoy what we have, but rather work to obtain more and more. In contrast, hunter-gatherers and people who live in subsistence cultures work to acquire only what they need, and do seem, by many measures, to be happier. David noticed this in traveling around the world with his band Talking Heads. "We would go to the outskirts of towns where we were playing in Latin America, or Africa, or Eastern Europe and see people who—compared to us—have very little in the way of material goods. Obviously no Wi-Fi, no air-conditioning, no electricity or refrigeration, but they live as they have lived for thousands of years, and they're happy. Even more noticeable is that there is a cohesion. Westerners like us feel like they don't have much, but they have something that I will probably never have: their social networks, their family, and their centeredness and rootedness."

Anthropologists note that all human societies look for God and for meaning, but the specific ways they do so vary enormously. What is recognizable from one age to another, and one culture to another, is the drive; what can be fascinating is the different ways that this uniquely human drive becomes channeled. We don't know if any animals have spiritual thoughts. Chimpanzees, dogs, and African gray parrots certainly behave differently when separated from those they love, something we might label despondency or depression. But if they have the ability to reflect on *why* they are experiencing the emotions they are—to realize "I sure would feel better if my owner Irene was here"—there is no evidence of that. They may well live in a world of the ever-advancing present, with no ability to plan, contemplate future or past, mourn, or look forward. A psychological study of dogs some years ago addressed the common experience of dog owners that their dog was there to greet them when they arrived home, suggesting to many (hopeful) dog lovers that their beloved anticipated their arrival, waiting by the door with a mental image of their imminent return. In fact, under controlled experimental conditions using hidden cameras, the dogs did not wait by the door, they simply were able to hear the car or footsteps of their owner a half block away and went to the door in what may have simply been an act of Pavlovian conditioning (hear car, go to door, owner comes in and makes a big fuss over me).

A wide range of animals use song across a diverse array of instances, but no animal has been observed composing or singing a song of longing, or love, or spiritual yearning. And yet all human groups do. The musical brain brought a new hum of neural activity between the brain's rational and emotional centers, along with all the billions of new connections possible with the enlarged prefrontal cortex. Self-consciousness and perspective-taking emerged and as far as we know are unique to humans. They lead most of us at some point in our lives to think about our place in the world, to think about the nature of thoughts, to pose questions and to look for answers.

Religion grew out of this desire to make sense of the world. Even without explicit training, most children reach a point where they ask: "Where did I come from?" "What was I before I was born?" "What happens when you die?" And, looking around at the world, "Who created all this?" Every human society that historians and anthropologists have uncovered has had some form of religion, and a belief system under which these questions are addressed. Some have even claimed that science is a religion, with its own rules of behavior and its own explanations about the origins of the world and of life, many of which rest on unobservables.

Much of what we know about the thoughts and beliefs of early humans is necessarily speculative, because they were not literate and did not leave us detailed explanations. Anthropologists make inferences, however, by visiting contemporary humans who live in societies that both lack written language and have been cut off from the rest of the world for thousands of years or more. These cultures tend to be composed of hunter-gatherer, pre-agricultural humans living in small groups. Their dominant belief is not that the world functions according to predictable, logical principles, but rather that events unfold at the whim of capricious gods who require various rituals or sacrifices in order to provide water, food, cure illness, and allow women to bear children. These beliefs are often based on a combination of superstition and lore handed down from generation to generation. A baby becomes very sick, a village elder sacrifices a wild boar, and the baby is cured. The next time a baby becomes sick, a wild boar can't be found and so the elder sacrifices a possum. The baby dies, and the elders come to believe that only boars can appease the gods. Hundreds of coincidences like this lead to rituals that form the basis of an early religion, based (typically) on pantheism, sacrifice, pleas, prayer, and appeasement.

One can argue that among the most significant events in all of human history was the invention of monotheism. Monotheism transformed the dominant worldview from one in which events happened for no apparent reason (at the whim of capricious gods) to one in which there existed a logic and order in things (according to the plan of the one true God). The laws of nature and natural processes were seen as the product of a rational, intelligent being. The advent of monotheism put an end to child sacrifice (which was ubiquitous in the pre-monotheistic world) and ushered in an era of logic. This swiftly led us to the Age of Reason, the Enlightenment, and science.

The cognitive capacity and drive toward holding religious/spiritual beliefs (though not necessarily the beliefs themselves) underlie the foundation of society, according to Rappaport. Human organization could not have come into existence in the absence of religious beliefs. Societies, by necessity, are built upon orderliness, organization, and cooperation. In many cooperative undertakings, such as building granaries, fending off invaders, plowing fields, providing irrigation, and establishing a social hierarchy, members of society must accept certain propositions as true, even if they are not directly verifiable. Preparing food in a certain way allows us to escape toxins in the food. A leader asserts that a neighboring tribe is planning to attack and we need to either prepare a defense or launch a preemptive strike. A wait-and-see approach is potentially calamitous—we need to act on faith.

Religions trained us and taught us to accept society-building, interpersonally bonding propositions. (Whether we still need religion in an age of science is a separate matter, and one that I don't want to get distracted by here.) Ceremonies with music reaffirm the propositions, and the music sticks in our heads, reminding us of what we believe and what we have agreed to. Music during ritual is designed, in most cases, to evoke a "religious experience," a peak experience, intensely emotional, the effects of which can last the rest of a person's life. Trance states can occur during these experiences, resulting in feelings of ecstasy and connectedness. Because the sacred belief is associated with the ecstatic state, it becomes reconfirmed in the experiencer's mind, with the music acting as an agent for reconformation every time it is played, ad infinitum. The emotion marks the belief. Three emotions in particular are associated with religious ecstasy: dependence, surrender, and love. These same three emotions are believed innately present in animals and human infants and were no doubt present in humans before religion gave them a system for expression and indeed for uplifting thoughts in self-conscious adults.

It is especially true that a cornerstone of contemporary society is trust and the ability to believe in things that are not readily apparent, such as abstract notions of justice, cooperation, and the sharing of resources implied by civilization. Indeed, modern technological civilization requires that we trust millions of things we cannot see. We have to trust that airline mechanics did their jobs in tightening all the bolts, that drivers on the road will keep a safe distance and stay within the lines, that food-processing plants observe health and hygiene codes. We simply cannot verify all these propositions directly—any more than the religious can verify the existence of God. The fundamental human ability to form societies based on trust, and to feel good about doing so (via judicious bursts of oxytocin and dopamine), is intimately linked to our religious past and spiritual present.

And music has been there to imprint these thoughts on our memory, sometimes long after a ritual or ceremony has ended, and long after an epiphany or revelation has passed. Music is able to do this because of its internal structure. Like human languages, human music is highly structured, organized, and hierarchical. Although the details of musical syntax remain to be worked out, there exist multiple redundant cues in music that constrain the possible notes that can occur in a well-formed melody. The human brain is an exquisitely sensitive change detector, and to be such, it has to register minute details of the physical environment in order to notice any violations of sameness, any deviations from the ordinary. The newest evidence, from the laboratories of Dick

Aslin and Elissa Newport at the University of Rochester, and of Jenny Saffran at the University of Wisconsin, Madison, shows that even human infants are sensitive to patterns and structure, detecting even minute changes in a musical sequence, and noticing when a sequence or chord progression is atypical.

The most surprising conclusion from this work is the *way* in which human infants accomplish this: Their brains (as do those of adults) compile statistical information about which notes are most likely to follow other notes (an ability afforded by the *rearrangement* or computational abilities of the musical brain). They do the same for language, learning a complex calculation of probabilistic regularities as to which speech sounds are most likely to follow which others. It is in this way that infants "bootstrap" a working knowledge of speech and music and a sophisticated awareness of what combinations are typical and which are atypical in the language and music to which they've been exposed.

What's exciting about this research is that it offers a parsimonious explanation for how both language and music are acquired. It also offers a compelling account of *why* music is so memorable, why we can still sing along with a song on the radio we haven't heard since we were fourteen years old, and why songs serve as such effective mnemonic devices for the knowledge of civilizations and the following of rituals and religious practice. The reason is because of the multiply embedded cues of melody and rhythm, constrained by form and style, as encoded in a series of statistical maps and, ultimately, statistical inferences.

The musical brain doesn't *have* to remember every note or every chord sequence; rather, it learns the *rules* by which notes and chord sequences are (typically, in a given culture) created. Violations of those rules are encoded as surprising events and so remembered as schema-breaking exceptions. We don't have to relearn every time a friend gives us his phone number that there are going to be seven digits plus an area code—this information is *schematized*. We don't need to learn that the candle-lighting song for a particular ceremony uses only certain notes in a certain pattern, because the choice of notes is constrained by the form (the scales) of our culture's music; we learn the exceptions and the rules, not every single note.

Music is therefore a highly efficient memory and information transmission system. We don't like it because it is beautiful, we find it beautiful because those early humans who made good use of it were those who were most likely to be successful at living and reproduction. We are all descended from ancestors who loved music and dance, storytelling, and spirituality. We are descended from ancestors who sealed mating rituals and wedding ceremonies with song, as we do now (or at least boomers like me) with "The Wedding Song (There Is Love)," the Carpenters' "Close to You," Nat King Cole's "Unforgettable," and Billy Joel's "Just the Way You Are." Songs like these remind us during life cycle events that we are part of a chain of continuing ceremony and ritual, participating as our ancestors did, binding our collective past to our personal future.

The Psalms of the Old Testament—reportedly written by King David—were all written as songs to commemorate, uphold, and celebrate the world's first monotheistic religion. The Catholic Mass, Handel's *Messiah,* the liturgical chants drawn from the Qur'an, and thousands upon thousands of other songs are intended to do the same. (Dan Dennett has suggested that atheists would do well to have some rousing pro-science gospel-like songs.) Some of the most beautiful music ever written has been songs of religion, songs of praise to God. Religious thoughts takes us outside ourselves, lift us up higher, elevate us from the mundane and day-to-day to consider our role in the world, the future of the world, the very nature of existence. The power of music to challenge the prediction centers in our prefrontal cortex, to simultaneously stimulate emotional centers in the limbic system and activate motor systems in our basal ganglia and cerebellum serves to tie an aesthetic knot around these different neurochemical states of our being, to unite our reptilian brain with our primate and human brain, to bind our thoughts to movement, memory, hopes, and desires.

Two final and important ways in which religious music has functioned in the formation of human nature are its ability to *motivate* repetitive action, and to bring what psychologists call *closure*. Obtaining closure ameliorates the very human tendency to obsess, to "stress out" over the unknown, to dwell on things that are beyond our immediate control. We pray for a sick child and then move on. We pass through a rite of passage and become in the eyes of society an adult. Rituals, religion, and music unite memory, motion, emotion, control over our environment, and ultimately feelings of personal security and safety, and agency. Some form of ritual is an integral part of the daily lives of children and adults in every culture. The sheer diversity of it is surprising, from people gathering sticks and bundling them in a certain way, to brushing one's hair one hundred times before going to sleep, to singing songs of praise upon arising in the morning, or whispering "I love you" to your partner before closing your eyes.

My mother's mother—the piano-playing grandma—followed a daily ritual of her own making after we bought her that electronic keyboard for her eightieth birthday. Every morning she woke up with a sense of purpose, a goal, and that was to sing her song, "God Bless America." Who said you can't teach an old grandmother new tricks? Learning the sequence of finger movements kept her mind active and challenged, especially as she began to add chords when she was eighty-nine. It gave her a sense of mastery, of accomplishment. And the particular song she chose to sing gave her pride in being alive and living in a free society. She played that on-awakening song every morning until she was ninety-six, adding in a personal prayer of thanks for her health, her family, her home, and her dog. Then one day she passed away.

I flew to Los Angeles for the funeral. She had a plot next to my grandfather Max's, way outside of town to the north. It was cold that morning and we could see our breath in puffs of steam rising toward the sky as the rabbi spoke the ancient prayers, the familiar cadences we had all known since childhood, the guttural sounds of Hebrew and Aramaic that reminded me of her own throaty German accent. I helped to carry her casket to the grave site, along with my father, uncles, and cousin Steven. The casket seemed too light to hold my grandmother, a woman of such determination, such strength and power that she had saved her entire family from the hands of the Nazis through the sheer force of her *will*. After the casket was lowered into the ground, we each picked up a fistful of dirt and, in the Jewish tradition, threw the damp earth into her grave. We sang from the Bible, Psalm 131, according to the ancient Aramaic tune that Jews have been singing some version of for two thousand years. The tune's Middle Eastern, minorish sound, with odd, exotic intervals, evokes stone buildings and walled cities:

Lord, my heart is not haughty, nor mine eyes lofty;
neither do I exercise myself in things too great,
or in things too wonderful for me.
Surely I have stilled and quieted my soul;
like a weaned child with its mother,
my soul is with me like a weaned child.
O Israel, hope in the Lord from this time forth and forever.

It was not the memorial speeches that brought us to tears, not the lowering of her casket into the ground, but the haunting strains of that hymn that broke through our stoic veneer and tapped those trapped feelings, pushed down deep beneath the surface of our daily lives; by the end of the song there wasn't a dry cheek among our group. It was this event that helped all of us accept the death of my grandmother, to mourn appropriately, and, ultimately, to replace rumination with resolution. Without music as a catalyst, as the Trojan horse that allowed access to our most private thoughts—and perhaps fears of our own mortality—the mourning would have been incomplete, the feelings would have stayed locked inside us, where they might have fermented and built up tension, finally exploding out of us at some distant time in the future and for no apparent reason. Grandma was gone; we had shared the realization and etched it in our minds, sealed with a song.

Many of the greatest music of all time has been religious—from the Song of Solomon to Handel's *Messiah,* to "Amazing Grace." Scientists and other religious skeptics often deride the religious by posing the following question: If God is so great that he created the entire universe, why would he care whether we praise him or not—why would such a powerful being be so psychologically needy that he wants us to sing to him?" But modern religious thinkers who believe in the existence of God indisputably suggest that the primary reason for this is to benefit the singer, not God. "God doesn't need our praise," Rabbi Hayyim Kassorla says. "He is not vain, he doesn't need us to tell Him He is great. But because He designed us, He knows what *we* need. He dictated that we should sing songs of religion and belief because He knows they help us to remember, they motivate us, and they bring us closer to Him; He knows that they are what we need."

CHAPTER 7

Love

or "Bring 'Em All In"

Romantic love songs are a sham that perpetuate a lie on unsuspecting young kids," said Frank Zappa. "I think one of the causes of bad mental health in the United States is that people have been raised on 'love lyrics.' "

Joni Mitchell says, "There's no such thing as romantic love. It was a myth invented in ancient Sumeria, repopularized in the Middle Ages, and one that is clearly not true. Romantic love is all about 'I' this and 'I' that. But true love is about 'other.' "

For two such different people to agree on the ultimate deception of romantic love—an avant-garde composer best known for perverse, cynical lyrics of the "Kenny picked his nose and left it on the window" type and one of the great romantic poets of our time—there must be something to the idea. Virginia Woolf described romantic love as "only an illusion. A story one makes up in one's mind about another person."

But what about all those shamelessly romantic love songs I loved so much when I was thirteen? The bubblegum songs?

Imagine me and you, I do
I think about you day and night, it's only right
To think about the girl you love and hold her tight
So happy together

I can't see me lovin' nobody but you
For all my life
When you're with me, baby the skies'll be blue
For all my life

There's Elvis's "Love Me Tender" and "(Let Me Be Your) Teddy Bear," Tommy Roe's "Dizzy" (and oh, what a great string part!), the Archies' "Sugar, Sugar" (sung by Ron Dante, who was back a few months later with "Tracy" by the Cuff Links), the Jackson Five's "I Want You Back," Gary Puckett and the Union Gap's "Over You," the Ohio Express's "Yummy Yummy Yummy." Not to mention great love songs from my parents' generation, such as "Our Love Is Here to Stay" as sung (in my favorite versions) by Ella Fitzgerald or by Nat King Cole:

It's very clear, our love is here to stay
Not for a year, but forever and a day

After these poetic statements about the writer's love in general, the lyric playfully turns to specifics to involve a metaphor for the longlastingness of their love:

In time the Rockies may tumble, Gibraltar may crumble
They're only made of clay, but our love is here to stay

Like many preteens, I "learned" about love from songs like this, from fairy tales and Disney movies. The message in these is that when you find the right person (and there is only one "right person" for each of us), you will know it's love because you will want to be with that person all the time, she will make you feel good, happy, and fulfilled, and you will never have a disagreement. In 1988, when I was working for Columbia Records, a friend of mine in the company played me an album that had just been finished by a new singer-songwriter with the audacious name of Parthenon Huxley. The first two lines of the second song on the album caught my attention: "I fell in love when I was twenty-one/I knew it was love, it was more fun than being alone." *That* is what love felt like to me—the discovery that there was no one in the world I would rather be with. The night before I got married, Julia Fordham dedicated a song to us at one of her concerts, saying it was "for a love that is filled with shiny, limitless newness." I sensed in her voice a world-weary knowledge of some transformation she thought would inevitably occur, but I didn't stop to figure out what she might have meant.

I knew about the curmudgeonly anti-love diatribes of some writers, and figured they were just trying to be funny. Kurt Vonnegut wrote:

> I have had some experiences with love, or think I have, anyway, although the ones I have liked best could easily be described as "common decency." I treated somebody well for a little while, or maybe even for a tremendously long time, and that person treated me well in turn. Love need not have had anything to do with it. Also: I cannot distinguish between the love I have for people and the love I have for dogs. When a child, and not watching comedians on film or listening to comedians on the radio, I used to spend a lot of

> time rolling around on rugs with uncritically affectionate dogs we had. And I still do a lot of that. The dogs become tired and confused and embarrassed long before I do. I could go on forever.

W. Somerset Maugham weighs in with "Love is what happens to a man and woman who don't know each other." According to the cold, clinical view of science, Zappa, Mitchell, and Vonnegut may be a lot closer to the truth than the Turtles, Ella Fitzgerald, or Parthenon Huxley, and Maugham may be the closest of all. Was love really here to stay or was what we were feeling just kid stuff? "And they called it puppy love." It may have been wonderful, powerful, and full of youthful energy, but not very mature.

Researchers have identified neurochemical changes that occur during the first few months of a relationship; huge releases of oxytocin (the "trust" hormone) and feel-good hormones like dopamine and norepinephrine, and at such high levels that they could be regarded as inducing clinically verifiable altered states of consciousness. The Stylistics crooned "I'm stone(d) in love with you," B. J. Thomas sang that he was "hooked on a feeling/high on believing/that you're in love with me." Bryan Ferry sang "Love is a drug and I need to score," and Robert Palmer sang "Doctor, doctor, give me the news/I got a bad case of lovin' you/No pills are gonna cure my ills." The Beatles harmonized "And when I touch you I feel happy inside/It's such a feeling that my love/I get high, I get high" (in Dylan's famous misunderstanding of the lyric, which was actually "I can't hide, I can't hide"). This neurochemical high causes our heart rate to speed up when we think about our loved one, impels us to make resolutions such as losing weight or exercising more, and fills us with a kind of giddy optimism that with this person, everything will work out right.

The drug aspect of love is reflected as more sinister in other songs, such as "Cupid's Got a Brand New Gun" by one of my favorite songwriters, Michael Penn (Sean's brother, and husband of Aimee Mann):

This quick opiate
might wear the wings of angels
that's when you realize
you've been shot down
wounded unto death by something called love

Penn suggests that love is a kind of death: a death of our single self, a death to some extent of our ego and of boundaries we place around our most private thoughts and feelings. The implicit message in his lyric is something we've all experienced, that love can make you do things you might not otherwise do, as Percy Sledge sang about in "When a Man Loves a Woman" (the lyric in Chapter 1). Having had romantic love and lost it can be one of the most painful things we experience—so much so that, like the person who drank too much alcohol the night before, we resolve never to do it again: "No, I don't want to fall in love (this love is only gonna break your heart)" (Chris Isaak); "I don't want to fall in love" (Tonya Mitchell); "I don't want to fall in love with the idea of love" (Sam Phillips).

"In songs like that," Sting says, "you really have an aspect of knowledge songs and love songs combined—they're trying to teach you about love, to be wary: 'Don't put your faith in love, my boy, my father said to me/I fear you'll find that love is like the lovely lemon tree/Lemon tree very pretty and the lemon flower is sweet/But the fruit of the poor lemon is impossible to eat.' "

The fact that romantic love can be reduced to or described by distinct changes in neurochemistry does not make it any less real. Stubbing your toe and winning the lottery also cause neurochemical changes, but that doesn't mean that when the brain chemistry is back to normal your toe isn't still bruised, or that your bank account isn't still flush.

At the historical core of this propensity for romantic love is the ability to form a strong partnership with another person, and that has clear evolutionary advantages—with the long maturation period of human children, those men and women who feel bonded and committed to one another are more likely to share in the raising of the children, and those children are more likely to thrive both physically and psychologically. Nowhere in your lineage, no matter how many thousands of years you go back, will you find an ancestor of yours who failed to have children. And although care may have differed considerably from one child to another and from one family unit to another, none of us has an ancestor who didn't receive at least the care necessary to grow up and successfully reproduce. Life in any era is unpredictable and child rearing is potentially fraught with difficulty. Feeling committed to your partner through romantic love confers obvious advantages to the offspring.

Unfortunately, the neurochemical high doesn't last forever. Sometimes it is gone after a few days or weeks or months; sometimes it can last five or seven years (leading to the so-called "seven year itch" in marriages). Perhaps the second most common song in pop music, after the romantic love song, is the breakup song, or the song of love lost. In "Let It Die," Dave Grohl of the Foo Fighters writes:

Hearts gone cold and hands are tied
Why'd you have to go and let it die?

In "Downbound Train," Bruce Springsteen sings:

She just said, "Joe, I gotta go
We had it once, we ain't got it anymore"

Or as Rosanne Cash sings, with the heartbreak dripping from her voice in her song "Paralyzed":

I picked up the phone, you were both on the line
Your words to each other froze me in time
A lifetime between us just burnt on the wires
Dissolved in a dial tone, consumed in your fires

What is this thing called love that it is so slippery, so ephemeral, so capricious? Could some of the greatest literature and music of all time have been written about an illusion? For those legions of modern thinkers and scientists who are atheists, there is certainly precedent in all the great writing and painting that was directed at a nonexistent God.

Because romantic love is what gets written about, talked about, filmed and sung about so much, we can temporarily forget that love comes in many different forms—the love between parents and children, between friends, love of God, love of one's way of life, and love of country. What all these forms of love have in common is intense *caring* (the opposite of love is not hate, but indifference), caring more about someone or something else than you care about yourself. Describing a man who played his last game of chess before committing suicide, Gabriel García Márquez writes in *Love in the Time of Cholera*: "Jeremiah de Saint-Amour, already lost in the mists of death, had moved his pieces without love." "Without love" is rendered as the equivalent of "without care." Parthenon Huxley writes in "Buddha Buddha":

> *Everything I do, Buddha did with love and that's what I aspire to,*
> *Try to rise above the petty things I run into*
> *I think of love when I do everything I do*

Love is about feeling that there is something bigger than just ourselves and our own worries and existence. Whether it is love of another person, of country, of God, of an idea, love is fundamentally an intense devotion to this notion that something is bigger than us. Love is ultimately larger than friendship, comfort, ceremony, knowledge, or joy. Indeed, as the Four Wise Ones once said, it may be all you need.

Romantic love is typically blind, as Maugham notes; we feel it for those we don't really know. And it tends to be very *me*-oriented: I love her because of the way I feel when I'm with her; because I have fun when I'm with her; because *I* find her beautiful, sexy, smart, funny, and so on. A more mature love comes when we care more about that person's happiness than we do our own—the selfless love that parents show toward children, a willingness to do without so that the child or our mate can do with. Romantic love drives us to be with the other person at all costs; mature love drives us to want to see the other person happy, even if that means not being with us. "If you love somebody, set them free," as Sting famously sang.

From an evolutionary perspective, it may seem that putting others before oneself doesn't make sense—after all, the name of the evolutionary game is to put your genes first. How in the world could *not* putting them first end up getting more of them in circulation? The apparent paradox of this kind of altruism has an explanation in evolutionary science. Because we share half our DNA with our siblings, sacrificing ourselves for a sibling or a sibling's offspring still helps some of our DNA, some of our genes, to survive. The same argument is made for homosexuality, which on the surface of things might seem to be maladaptive. But if a gay brother or sister cares for a sibling's children, he or she is still helping to promote the family's genes. Altruism also helps to diffuse potentially deadly conflicts. During harvest festivals celebrated by Southwest American Indians, for example, foodstuffs are redistributed equitably among neighboring tribes, eliminating what could be deadly food-jealousy wars.

Altruism is not limited to humans—animals show it as well. Vervet monkeys delivering alarm calls to warn others are putting themselves at risk (by attracting the attention of the predator) in the service of protecting their kin. Dolphins have been shown to aid other species who are injured by helping them to get to the water's surface or to shore.

Some evolutionary biologists argue that love developed as an adaptation that helped increase the likelihood that human offspring would receive the care they require. Humans have the longest maturation period of any animal. Baby mice at three weeks can be left completely alone by the parents and survive. By twelve weeks dogs are self-sufficient. But leave a nine-month-old human infant alone and—even if no harm comes to the child—you can be arrested for child endangerment. Human children need at least a decade of care and instruction. Unlike spiders, bees, and birds, who have instructions encoded in their brains for building their webs, hives, and nests, human infants learn by being taught explicitly. Using the same phrase that Joni Mitchell used to describe humans, the anthropologist Terrence Deacon has referred to this as *de*-evolution, in that the brain itself carries fewer and fewer preprogrammed instructions (compared to other primates and other mammals), and culture and experience take on a greater role in shaping education and behavior. This appears to be related to humans' fantastic adaptability and ability to thrive in disparate environments, far more so than apes and monkeys. The band Devo (named for the principle of de-evolution) satirically wrote in their song "Jocko Homo": "God made man, but he used the monkey to do it." Or as XTC sang, "We're the smartest monkeys."

Across all species, young brains are more sensitive to environmental input and more resilient in recovering from brain injury than older brains. This reflects a parsimony on the part of evolution: Rather than building into genes and brains information that is ubiquitous and readily available in the environment, brains are configured such that they can incorporate regularities in the environment, learning through exposure. They do this through the initial *over*production of neurons that are later selectively pruned (in a kind of Darwinian selection process). The system is thus designed to configure itself. It needs to, because as we grow, our brains need to adapt. For example, as we grow taller and heavier, we need to adjust the force we apply to walk. As our eyes grow farther apart (because our head gets bigger), we still need to reach and grasp, and so the brain must take into account these differences in the binocular disparity. If we were designing the brain for efficiency, we would create a system that can learn rules, can map itself, and can respond to the particulars of the environmental input it receives. This efficient parsimony also confers flexibility in the case of unusual circumstances. For example, an organism born with one functioning eye instead of two maps all the input from that one eye to the entire visual cortex (including regions normally reserved for projections from the other eye)—this way the cortical real estate for the nonworking eye isn't sitting empty.

The brain learns music and language because it is configured to acquire rules about how musical and linguistic elements are combined; its computational circuits (in the prefrontal cortex) "know" rules about hierarchical organization and are primed to receive musical and linguistic input during the early years of development. This is why a child who is denied exposure to music or language before a certain critical age (believed to be somewhere between eight and twelve years old) will *never* acquire normal music or language skills—the pruning process has already begun, and those neural circuits that were waiting to be activated become eliminated. Certain universals of music suggest that the innate structures themselves contain loose constraints for how music will be represented. Among these are the octave, the fact that all musics work with a set of discrete pitches, and the ubiquity of simple rhythmic ratios (the durations of musical notes across disparate styles and cultures tend to occur in ratios of 2:1, 3:1, or 4:1, not more complex ratios such as 17:11).

The enlarged role of education of the human child stands in stark contrast to any other species. And across history, song has been one of the primary ways in which life lessons are taught. Our ancestors discovered that well-formed songs, combining musical and rhythmic redundancies with lyric messages, facilitate both the encoding and transmission of important information—knowledge songs. But it was love songs and the feelings of love that created the social structure in which we bring up children. Men and women form pair-bonds to the lilts of love songs and mutually ensure the care and nurturing of children.

Humanlike pair-bonding and monogamy are rare in the animal kingdom. In the vast majority of the 4,300 species of mammals in the world, adult males and females tend to be solitary, coming together only to copulate; males don't pair-bond with the mothers of their offspring and they don't provide paternal care. Even in the most social mammals, such as apes, lions, wolves, and dogs, there is no evidence that males even recognize their own offspring.

In humans, although polygyny (long-term simultaneous sexual relationships between one man and two or more women) has existed as a rare behavior for thousands of years, the dominant mode of relationships is of monogamy, or at least serial monogamy. This requires that we establish bonds and feelings of intense attachment; *love* and its neurochemical correlates can be seen as the evolutionary adaptation that makes these long-term bonds possible. Once the adult male-female love mechanism is in place, it can be easily adapted for parent-child love. In fact, Ian Cross quips, love serves an important function there—human infants can be noisy, fussy, and a lot of trouble, and love for them may be the only thing that prevents many parents from killing their children.

Love and altruism take on a different quality in humans than in animals (as do many other of our behaviors) because of our awareness of them, our self-consciousness. We can plan how we want to demonstrate our love, we can promise to love. Our perspective-taking ability helps us to recognize that we have to win over the skepticism of our potential mate.

Some readers may object that when considering the survival of the species, and evolutionary adaptations, love does not appear all that important compared to some of the other attributes we've seen in *The World in Six Songs*. For example, the drive toward knowledge seems clearly essential—those individuals who enjoyed learning were better able to adapt to changing environmental and social conditions. (And as a consequence, they were favored by natural selection.) Knowledge songs developed as an efficient way to encode, preserve, and transmit information. As early or protohumans left the shelter of the trees for the open savannah, exposing themselves to predators, the drive toward friendship allowed for us to navigate complex social and interpersonal exchanges. Comfort songs helped to reassure infants and others with whom we weren't in physical contact that we were nearby, and they helped to pick us out of periods of sadness by reminding us that others too had felt sad and recovered.

Joy songs began as expressions of our own emotional states, signaling to those around us either a positive outlook or the possession of food and shelter resources. Neurochemical boosts associated with joyful singing helped to reinforce joy as a signal for mate selection. Religion and its songs served to bind animal rituals into systems of belief, and ultimately helped to systematize and socialize feelings of hope and faith.

Love, as intensely as we feel it today, and as much attention as it receives in popular culture, art, and daily conversation, would seem the least important of these, a titillating but nonadaptive neurochemical high similar to the one we get from cocaine, marijuana, a fine Château Margaux, or a good double espresso. If love is viewed only narrowly as romantic love, then it is probably not a cornerstone in the creation of human nature. But love in its larger sense—the sweeping, selfless commitment to another person, group, or idea—is the most important cornerstone of a civilized society. It may not have been important for the survival of our species as hunter-gatherers and nomads, but it was essential for the establishment of what we think of today as human society, what we regard as our fundamental nature. Love of others and of ideals allowed for the creation of systems of courts, justice that is meted out to all members of society equally (without regard to financial status or race), welfare for the poor, education. These fixtures of contemporary society are expensive in terms of time and resources; they work because we believe in them, and are willing to give up personal gain to support them.

I mentioned "I Walk the Line" as an example of a knowledge song, because the singer is reminding himself to "toe the line," to be true. But of course, it is also a powerful love song, a song celebrating a commitment to something above the passing emotions of lust.

Our love for another person, that special someone, a single love interest, pulls us out of ourselves and lifts our thoughts to a grander scale: How can I make the world better for this person? When I was in my twenties, the only love I experienced was the immature, selfish love of "I love her because she makes me feel good. I want to make her happy so that she'll stay with me." Now, at fifty, I think about the woman I love in terms of what *she* wants. I want to make her happy because I can't be happy when she is unhappy. We discover that the act of giving love is more powerful than getting the hug you need—if we can get over our own hunger for love, then we have reached the state of pure love, of being connected to a larger ideal, bigger than our own individual life.

Love songs, like all art, help us to articulate our feelings. They often use metaphorical language ("I'm on fire," "I will climb the highest mountain") to help us see our emotions from a different perspective. They stick in our heads to remind us, as the emotions ebb and flow, of what we once felt. And above all, they raise the feelings to the level of artistic expression—imbuing them with an elegance and sophistication that helps us strive for them even when the going gets tough.

To understand where love songs came from, it is necessary to go back in evolutionary history and ask two questions. First, of all the senses that could do this work, why does *sound* have such an important role in our emotions (or, in other words, what are the evolutionary origins of hearing and music)? Second, how did the evolutionary changes that gave us the musical brain give us the sort of consciousness that is required to compose songs, to create art and science, and to build functioning societies?

The hair cells that we have in our ears are found in all vertebrates, including fish, and are structurally and functionally similar to those found on the legs and bodies of many insects (where they are called *sensilla*). When a grasshopper moves its leg, its hair cells are stretched and help to indicate the position and location of the leg. They are also sensitive to air, water, and other currents, to help detect the presence of an object approaching. This points to the phylogenetically early use of hair cells not just for detecting changes in pressure, which led to hearing in mammals and fish, but changes of position, which led to the vestibular system, our sense of balance. Hair cells are so sensitive that a stretching or movement of only 100 picometers causes them to fire—that's 1/100,000,000 millimeters, or 100,000 times smaller than a chromosome and 10 times smaller than the radius of a hydrogen atom.

The eardrum is a thin membrane stretched out taut inside our ears, and changes in pressure—whether in air, water, or another medium—cause it to wiggle in and out. This pattern of wiggling eventually sends signals to a snail-shaped organ in the inner ear called the cochlea, which is lined with hair cells much like insects' sensilla. The human cochlea is so sensitive that it can detect vibration as small as the diameter of an atom (0.3 nm) and it can resolve time intervals down to 10μs—if a sound ten feet away from you moves even two-and-a-half inches to one side, you can detect that movement just by the difference in the sound's arrival time at your two ears. The ear detects energy levels a hundredfold lower than the energy of a single photon. Hearing is so sensitive that some species can hear the footsteps of the insects that they seek to eat.

The advantage of hearing over other senses, such as vision for example (as I mentioned in Chapter 2) is that sound transmits in the dark, travels around corners, and can reach us when there are visual obstacles between us and what we want to hear. Sound constitutes an effective early warning system for something approaching us—a boulder rolling uncontrollably down a hillside, a predator stepping on a twig outside our cave, and so on. As part of the early warning system, our hearing sense also has immediate neural connections to our startle response, and detects even the slightest change in background noise in the environment.

Evolution might well have found *other* ways for us to gather information about the environment rather than the senses we know. Indeed, some animals employ systems that are exotic compared to ours. Sharks have an *electrical* sense—a sensory system that detects electrical fields produced by the neuromuscular activity of potential prey. Bees, ants, turtles, salmon, sharks, and whales use a magnetic sense for orientation. Indigo buntings possess a celestial compass that allows them to fly at night to find north; through evolution, they have internalized the fact that the entire sky revolves around Polaris, and so they navigate based on the one star that doesn't change position in the night sky. Interestingly, bunting genes don't specify which star is the North Star, only that the invariant star should be treated as north (allowing for the possibility that buntings could navigate throughout the northern hemisphere without having to develop a separate mechanism for different latitudes). Experiments by Stephen Emlen with indigo buntings in a planetarium demonstrated that the birds will treat any star as the reference point if it stays stationary.

Given that evolution gave all vertebrates a sense of hearing, it isn't obvious that this would develop into something as complex as music, but evolution moves slowly. Complexity is built up stone by genetic stone through small adaptations, each in itself perhaps imperceptible, building to a grand crescendo. As hearing became refined, and responsive to environmental events, selection pressures made all vertebrate brains sensitive to differences in pitch, spatial location, loudness, timbre, and rhythm, the fundamental ways in which objects can be differentiated from one another through sound. This is not so surprising, because the basic structure of neurons and synapses, the chemical soup of neurotransmitters, is common to all vertebrates.

The basic function and structure of genes is also common to all animals. Genes serve to determine, constrain, and guide cells so that they develop properly and perform their essential functions; they contain instructions, like a blueprint, that neurons and other cells follow. As fetal and infant brains develop, certain common proteins encoded in DNA, such as netrins and homoetic gene products, even dictate how neurons will connect with one another along specific pathways that are analogous in animals as different as roundworms, insects, birds, and mammals. The genetic instructions for neural development are both so powerful and so flexible that they can even guide neural connections to the right places when part of one brain is transplanted into another. Evan Balaban removed the auditory cortices from Japanese quail embryos and surgically implanted them into the brains of embryonic chickens. Not only do the grafts link up anatomically with the new host brains, but the host birds act in ways that demonstrate they have incorporated the donors' hardwired propensities—specifically, the chickens make vocalizations like quails, not like chickens, even when they are raised by other chickens.

Auditory pathways that are comparable to ours exist even in reptiles and birds. One interesting similarity of auditory system architecture is *tonotopy.* This means that frequency-selective neurons in the auditory cortex are configured so that low notes activate one end of it and high notes the other—the cortex is literally laid out like a piano keyboard! Tonotopy has been seen in the guinea pig, squirrel, opossum, ferret, tree shrew, marmoset, owl monkey, macaque monkey, rabbit, cat, and bush baby, plus many reptiles and birds. But even though these animals and humans share such tonotopic organization, there remains disagreement among researchers about animals' ability to differentiate pitches. It is clear that they can distinguish low tones from high tones, but as tones get closer together, it seems that many animals don't possess the same resolution that humans do: Three consecutive tones in our musical scale may all sound like the same tone to a marmoset, frog, or carp.

In all these species, however, the ability to locate a sound in space is highly evolved—neurons take projections from the two ears to indicate where the sound is coming from. The reason that stereo sounds so much better to most of us than mono isn't just that different sounds come from two different places instead of one, but that evolution favored those species who developed and used stereophony for sound localization; we like stereo because we are descended from ancestors who exhibited a selective advantage for this form of spatial processing, one that helped us to better locate (and escape from) predators.

Differences across species in several brain regions (particularly in the thalamus and auditory cortex) led to differences in the ability to remember sounds and their locations. It takes many, many more trials for a rat to learn an association between a sound and an event (such as a source of food or danger) than it does for a cat; primates are even faster learners. Another difference is that the higher up one goes, so to speak, in the phylogenetic ladder, the longer are the latencies for neural firing and the lower are spontaneous discharges from auditory neurons. In other words, more advanced species are *less* likely to startle. This makes sense because we humans rely less on sound-by-itself to make sense of the world than do lower animals. We combine sonic information with information from other senses, with memories, and also expectations about what is going to occur. Expectational processing reaches its peak in humans—we can prepare ourselves for a loud noise as we see a pin approaching a balloon; the binturong and the baboon are more likely to be bothered by the burst, no matter how many times they see pin pop the balloon.

I said earlier that I believe development and mutations in the prefrontal cortex created the brain structures that underpin the musical brain, which in turn allowed for the type of mental structures that were required in order for us to develop societies. Although the sense of hearing is shared across all vertebrates, separate changes in brain structure allowed many species to use the sense of hearing for communication with one another. Vocalizations, whether the croaking of frogs, the chirping of birds, or the pan-hoots of chimpanzees, served to signal physical and emotional states among members of a species, who developed brain mechanisms to both produce and interpret those signals. Of course, there is a risk in making any sound, because predators can more easily locate the sound maker; the evolutionary advantages of being able to communicate through sound had to outweigh the disadvantage of attracting predators.

Vocalization facilitates the process of sharing information, in turn enhancing survival. There is a very high correlation between the amount of vocalization and the closeness of social relationships in a species. In particular, pair-living species tend to vocalize more to attract mates, enhance pair-bonding, defend their families' territory and resources, and locate one another (especially in the dark or around corners or other visual barriers). In birds, 90 percent of the species live in pairs, and birds of course are famous for their vocal behaviors. In pair-living primates, like the siamang gibbon, owl monkey, and titi monkey, vocalization is a conspicuous feature.

During a typical day, chimpanzees—our closest relatives—associate in temporary parties that vary in size and membership, much as humans do. They may become separated from friends and family and have a strong need to be assured of future meetings and cooperating. Sound communication gives this reassurance. Because primatologists can identify an individual chimp by its call, we assume other chimps can too. Thus, rudimentary vocal communication may have developed as part of the increasingly complex social lives that primates lived.

As they left the cover of trees out on the savannah, protohumans exposed themselves to increased danger from predators. They needed greater brain power to help them to stay one step ahead of those predators and to deal with the wide range of environmental variation that came from their nomadic lifestyle. Diet plays a surprising part in the brain size story too, as biologists have found an inverse relationship between brain size and digestive tract size (which in turn is inversely related to food complexity). Those primates who eat leaves tend to have smaller brains but larger digestive tracts than those who eat fruit. The reason is that leaves are more difficult to digest than fruit, requiring more processing stages and more energy to break down the complex carbohydrates into usable sugars.

On the other hand, fruit-eating requires more cognitive skill, specifically the ability to remember the location of fruit-bearing trees, anticipate when they will be in season, and discern ripe from unripe (or rotten) fruit. This latter process benefits from improved color vision—the color of a fruit's skin carries information about how ripe it is, and therefore the nutritional content and digestibility of the fruit—all which require increased brain size in the occipital cortex. The total amount of energy available to an organism is limited, forcing an evolutionary trade-off between brain size and digestive tract size. Geneticists have found that humans have lost an ability possessed by most mammals to create vitamin C internally; the gene for L-gulonolactone oxidase (GULO) on chromosome 8 is nonfunctioning (defective) in humans and other primates, and this is believed to be a result of our becoming fruit eaters some 40 million years ago. Because vitamin C was available exogenously, we and our primate cousins didn't need to manufacture it anymore, and so the ability was lost through genetic drift. This is another example of Deacon's notion of parsimony: Evolution often "offloads" instructions or survival plans from the genome to the environment.

There are relatively few species with big brains like ours, and this is because big brains carry with them a high biological price tag: They are metabolically expensive in terms of the energy needed to furnish them with oxygen, to cool and to protect them. Complex brains take longer to mature and to train, and large-brained animals are correspondingly more dependent on their parents or caregivers and for a longer time. This means extra energy from parents, which in turn means fewer offspring, fewer opportunities for a parent to successfully pass on his or her genes to all of posterity. All these high costs must be outweighed by the benefits (or evolution would not have selected them), but the cost-benefit ratios are unlikely to be the same across species.

Our brains are not only large for our body size, but the prefrontal cortex, seat of the musical brain, is large compared to the rest of our brains. This region—just behind your forehead—is most highly developed in humans and tends to be largest in those species that are most social, for example chimpanzees, bonobos, and baboons. Well-connected female baboons have more babies who receive better care as the community chips in to help these well-liked, high-status females. Baboon social order, as represented in the prefrontal cortex, has an evolutionary basis.

When people think about other people, it is the prefrontal cortex that is activated. The connection between social behavior, communication, and the prefrontal cortex is strengthed by the fact that the cortex developed independently in a completely different biological order, *Carnivora*. As the zoologist Kay Holekamp recently discovered, the spotted hyena—an especially social mammal—also has an enlarged prefrontal cortex. "Spotted hyenas live in a society just as large and complex as a baboon's," she says. Because hyenas and primates last shared a common ancestor 100 million years ago, we infer that similar evolutionary forces worked independently to arrive at a similar adaptive solution: the prefrontal cortex as the seat of sociability. In humans, it evolved further to become the seat of music, language, science, art, and ultimately—society.

Clearly there are advantages to having the kind of brain we humans do, with enlarged, highly developed, intricately connected prefrontal cortices. So, you might ask, if that big prefrontal cortex is so useful, why don't *all* animals have one? But evolution doesn't work that way. We might just as well ask why humans don't all have long giraffe-like necks, and fishlike gills, and night vision like owls. Evolution selects adaptations that solve specific problems (and it has to build on existing structures). Each adaptation comes with certain metabolic costs, and only pervades the population when its advantages outweigh those costs. The *handiness* of being able to avoid using ladders or breathing underwater simply isn't the same as a biological necessity; convenience is an inadequate motivation for natural selection. We have big brains because they solved a specific problem. Typically such a solution is required when there is competition for food resources, or a need to escape environmental or predatory dangers.

Chief among these benefits would have been a greater ability to adapt to environments and to shape parts of the environment to meet our needs. Tool use is one important milestone in cognitive evolution. Archaeologists—especially cognitively oriented ones who are interested in the evolution of mind—talk a lot about stone tools, sharp flakes chipped from parent "cores," found in some early human excavation sites. Why all the fuss about some old rocks? It's because tool *making* as opposed to mere tool *use* (which crows and monkeys do) represented a major cognitive leap: It required a type of thought not before seen in any other species. These stone tools were the first implements that conformed to a "mental template," an *idea* in the mind of the beholder that existed prior to the completion of the tool. Stone tools are thus the first evidence we have of the birth of *symbolic thought,* a qualitative change in ability, one that distinguishes humans from other species and makes possible art and music.

Archaeologist Nicholas Conard discovered a mammoth tusk from the Ice Age—about 37,000 years ago—at a dig in southern Germany. Its existence suggests that humans must have brought musical instruments with them when they left Africa for Europe. The tusk had been split down the middle, hollowed out, and had holes made in it to turn it into a flute—all this would have required a great deal of craftsmanship, time, effort, and most importantly, a mental template of what the finished artifact needed to look like. As Ian Cross notes, "One of the most technologically advanced tools of the time was a musical instrument!"

The earliest known *Homo sapiens* fossils in Europe date to about 40,000 years ago. These European ancestors migrated from Africa and possessed not only the ability to fashion stone tools, but workmanship that displayed what the anthropologist Ian Tattersall described as "an exquisite sensitivity to the properties" of these materials. They brought with them carvings, engravings, and cave paintings; they kept their own history on bone and stone plaques; and they made music on flutes that they built from wood and bone. In short, migratory, preindustrial humans of 40,000 years ago had art and an artistic sensibility. "Clearly," Tattersall writes, "these people were *us*." The artistic remnants that have been left behind—carvings and paintings—show such sophistication and power that it is likely that they were not our ancestors' first attempts at art. Rather, they are the lucky ones that survived, and there clearly must have been a great number of refinements and improvements that led to these. Art, in other words, must have existed for tens of thousands of years *before* the earliest artifacts we've found.

The three cognitive components of the musical brain are perspective-taking, representation, and rearrangement. Perspective-taking encompasses the ability to think about our own thoughts (what some call metacognition or self-consciousness). That is, the ability to examine the contents of our own minds, hold them up to the light of day, to the light of reason and objectivity. It also entails recognizing that others have beliefs, intentions, desires, knowledge, and feelings that may be very different from our own. *I* may be feeling happy right now, but that doesn't mean that *you* are; I may know where the food is hidden and you may not. I sing a song to tell you what I am experiencing, as a way of bridging the separation of our minds, because I know that you do not necessarily experience what I do.

Representation is a cognitive operation that allows for displacement in time and space—thinking about things that aren't there now. I can talk about fear without becoming afraid; I can sing about sorrow that I don't necessarily feel right now. I can represent love with a ♥ or an arbitrary string of vocal utterances such as "luv," "amoor," or "aijou." Such symbolic representations constitute abstractions, and lay the foundation for the creation of visual (and other) art.

Rearrangement is an ability to combine and recombine objects in different ways, to organize them in theoretically motivated hierarchies, categories, to impose structure on objects based on shifting notions of their content. For example, given a list of words—*banana, baseball, grape, golf*—you could organize them into two groups, fruits versus sports. Or into two other groups, words that begin with *b* versus words that begin with *g*. Or three groups based on the number of syllables each word has. Rearrangement requires computational structures in the prefrontal cortex that other animals may have, but that only humans have learned to fully exploit. The three of these (perspective-taking, representation, and rearrangement) may have evolved independently, but together they are the foundations of the musical/artistic brain.

A number of distinct cognitive operations are necessary for the creation of art. Specifically, one needs to (1) form a mental image of the thing to be created; (2) hold that mental image in mind; (3) understand how to go about manipulating objects in the physical world in order to conform to the mental image; (4) compare the ongoing development of physical-world object with the mental image in real time; (5) update plans as necessary to accommodate unforeseen difficulties or mistakes in manipulating the physical object. As the old joke goes, to sculpt a bear you just start with a piece of rock and chip away everything that doesn't look like a bear.

But of course this is not trivial. Our cave ancestor might have tried to draw a bear with a piece of coal on a wall. First, he needed to understand that the drawing can never look exactly like the thing it represents—it is a *version,* an *abstraction,* of the real thing, an imperfect *approximation* of a mental image. That way of thinking requires the objectivity of perspective-taking—the ability to think about one's own thought process, one's limitations, one's relationship with the world. The artist needed to decide how to draw in a way that preserved essential, recognizable details. This selection process requires abstract (or symbolic) thought. Having drawn a few lines, he would have had to assess the creation objectively: Does this look like I *thought* it would when I started drawing it? This requires an iterative process, changing some aspects of the physical drawing to match the mental image. Finally he needed to ask himself: If someone else were to look at it, would *they* know it was a bear? This also requires the objectivity of perspective-taking, specifically the ability to recognize that other people have their own knowledge, thoughts, and beliefs that are not necessarily the same as our own.

Imagine now what is required in building and playing a flute. One needs to have at least an intuitive and practical (if not a reasoned and scientific) understanding that carving holes in a bone will allow for a change in pitch. One presumably has notes in mind before blowing, and if the flute sounds different from what is in mind, one plays around, experiments, iterates, forming some kind of convergence between the mental image and physical reality. This is of course what composers do, even the best of them, when they try out their ideas on instruments. Notwithstanding stories of Mozart and Beethoven composing entirely in their heads, the vast majority of pieces written by the vast majority of composers involved a "trying out" phase in the real world, an iterative process in which the physical and mental images of sound were brought closer together.

Indeed, many composers (like other artists) spend a great deal of time trying to match or approach some mental image, each new piece an experiment that brings them closer. If they're not successful or otherwise unhappy with the outcome, they keep on trying. Think of Van Gogh's series of paintings of sunflowers—why keep on painting sunflowers unless you are trying to perfect something about your representation of them? Paul Simon describes this process in music as an aesthetic goal that he approaches using a tool kit, a palette of musical ideas and techniques that he draws on to bring him closer to what he hears in his head.

"One of the main things that you have to decide when you make a record," Paul says, "is what's the *sound* you're going to make on that record. And in a larger sense you have to be able to recognize what are the sounds that you like. We all have access to pools of sounds, clusters of sounds, your personal tool kit. They're based on what you remember from a lifetime of music listening . . . what it is that you loved and collected in your mind as sounds that you like and then you go for those sounds all the time. Sometimes you don't even *like* the sounds, but you're stuck with them—take my voice. Sometimes I wish that that wasn't the voice that was singing the song, but that is my voice, you know. I'm not going to cover it up or anything; sometimes it's really appropriate to what I'm singing, sometimes it's inappropriate, and then I wish it could be somebody else's voice. 'Bridge Over Troubled Water' is an example of where I don't have the voice I wanted, so of course I got Artie to sing it. But if I could have had any voice, it would have been harder, more powerful, more like Otis Redding."

This experimenting and iterating toward a specific aesthetic goal shows up in one way with Paul's longstanding interest in polyrhythms and indigenous musics. He first explored these in 1970 with "El Condor Pasa" and "Cecilia," further developed them on "Me and Julio Down By the Schoolyard" and reached his artistic peak with them on a trilogy of albums, *Graceland, The Rhythm of the Saints,* and *Songs from The Capeman.* Similarly, Paul McCartney seemed to be trying to capture both the sound and the aesthetic essence of a forties dance-hall tune in a string of songs beginning with "When I'm Sixty-Four" (written in 1958, recorded in 1967), "Your Mother Should Know" (1967), and "Honey Pie" (1968). With each one, he got a little closer, until 1976, when he released "You Gave Me the Answer," with production and orchestration sounding almost exactly like a Fred Astaire record. McCartney never attempted a dance hall–style song after this, and so I assume that he finally met his artistic goal and moved on to other experiments and other challenges.

Interestingly, this cognitive leap that brought about art didn't follow a *sudden* change in brain size. All three features of the musical brain—perspective taking, representation, and rearrangement—arise from modes of thought that are surprisingly similar in other animals, and differences in brain anatomy that are quantifiable but not especially dramatic. This is because human mental abilities are built from existing structures found in other animals that show many similar features. In other words, there exists a continuum of mental skills in the animal kingdom, and in virtually every case of an ability, it is not that we humans have a wholly *different* ability (a difference of kind), rather, it is that we have *more* ability (a difference of degree). Darwin himself noted this.

How and why perspective taking, representation, and rearrangement came about in human brains we still don't know, but the proximal cause of the difference was probably unexceptional. In other words, small adaptations in the prefrontal cortex, perhaps too small to detect from the fossil record, allowed neural circuitry to cross a threshold of complexity that bestowed on our lucky ancestors the brain power to make these significant cognitive leaps. But the unexceptionalness of the biological change (the cognitive result is quite exceptional) can be understood by appreciating how much of what we can do can also be done by other species.

Take language communication as an example. For years it has been argued that many species have forms of communication, but only humans have *language*. Scientists state very specific demands for what constitutes language and what doesn't, and on the surface, the way that we humans communicate with one another does exhibit several key features that are different from what animals do. Among those features, there are no reported cases of animals *naming* things spontaneously. Some chimps and apes have succeeded in learning sign language, and dogs can learn to label objects differentially (my dog Shadow distinguishes several different toys by name including his "fuzzy man" from his "Cat in the Hat"), but this only demonstrates their ability to *link* a visual or acoustic stimulus with an object. There is no evidence that these animals understand that the name *refers* to the thing (that it has intensionality), only that they have associated the sign or sound and the object in a kind of Pavlovian, unconscious connection.

Moreover, once they've learned to name things, animals don't teach other animals to name things.

Another feature of human language is that it can be used reflexively. We talk about our language, discuss whether we're using a word appropriately, we make a distinction between whether we're relaying the gist of a conversation or quoting a person verbatim. We make "quote" symbols in the air with our fingers to indicate when we're using a phrase literally or ironically, and we use the *music* of language—prosody—to help distinguish such literalness from irony.

A third important quality of human language is that it is infinitely expandable—every day humans produce or hear utterances that have never before been spoken, and yet we understand them. Human language is expandable in two ways. First, there is no such thing as the longest sentence in the English language (or any other language we know of), because whatever sentence you nominate for the longest, I can always make it longer by adding a clause to the front of it, such as "David Bowie thinks that . . ." So "My dog has fleas" becomes *"David Bowie thinks that* my dog has fleas." Or the longish sentence, "The album *Brothers in Arms* by Mark Knopfler and Dire Straits, in addition to containing great songwriting and performances, may well be one of the most perfectly engineered and mixed albums in the history of recorded sound" becomes *"David Bowie thinks that* the album *Brothers in Arms* . . ." (and so on). In this respect, human language is like the natural numbers—there is no number larger than all the others, because I can always add the number 1 (or any other positive number) to the number you nominate as the largest. (In fact, mathematicians have found utility in talking about the difference between ∞ and $\infty + 1$.)

The other way that human languages are expandable is that

words can be combined in different ways. No child learns a list of all the possible sentences he will need; rather, he learns words, and then rules about how new words are formed and about how they can be combined. This comes under the musical brain's capacity for abstraction—we know that the words are merely elements of an utterance that *stand* for specific things, and that they can be recombined or substituted to change the meanings of utterances. "The cat chased the dog" means something different than "The dog chased the cat" by virtue of the ordering of the words. (Grammar specifies, among other things, rules about who is doing what to whom based on where in the sentence the elements appear.) We can say "I am going to the park" or "I am going to the zoo" using similar form and structure, but different words.

Animals use fixed, irreducible phrases. The black-capped chickadee of North America has thirteen distinct vocalizations, but there is no evidence that they can be combined, or that part of one can be substituted or inserted in another. The "chick-a-dee-dee-dee" call *is* the whole message. "Chick-a-dee-dee-doo" or "Chuck-ee-dee-dee-dee" are not possible messages in chickadee language.

Steven Mithen argues in his book *The Singing Neanderthals* that Neanderthals may have communicated using a kind of protomusic with pitch, rhythm, timbral, and loudness variations. Moreover, Mithen believes they employed a repertoire of fixed calls, constituting something slightly more sophisticated than those in use by crows and vervet monkeys, for communicating such holistic thoughts as "Look out! There's a snake!" and "Come get some food." Unlike human languages, in which words can be substituted in sentences to alter their meanings ("Come get some water"), the holistic Neanderthal utterances would have been more like monkey or bird calls—fixed and not extendable or changeable. They were neither language nor music as we think of them today. Darwin believed that music as we know it is a kind of fossil, a remnant of an earlier communication system or "musical protolanguage."

Noam Chomsky, the Hunter S. Thompson of linguists, proposed that language may be decomposable into two components—*conveyance* and *computation*. The distinction maps to two different, historically sequential forms of language. The first kind to appear was an unstructured protolanguage that conveyed concepts, meaning, and emotions, like Mithen's Neanderthal language, and like contemporary chimpanzee, monkey, and gibbon communication (and presumably like that of Australopithecines). The *conveyance* form of language served to communicate the *right-now*: the emotional state of the communicator, the existence of a predator or of food, and so on. The second kind arose later, when the human brain evolved computational modules capable of rearranging elements to a preconceived plan, and of imposing a complex hierarchy on objects, including linguistic objects; in the case of utterances, this constitutes a structure and order we call syntax. (I prefer the term *rearrangement* to the Chomsky's *computation*, and so I'll use that here.) This ability to understand, form, and analyze hierarchies is a key development that led both to music and to language. Chomsky doesn't say so explicitly, but it seems that along with this hierarchical processing came the ability to communicate about things beyond the right-now—to talk about things that happened in the past or might happen in the future.

I believe that Chomsky's conveyance and computational (rearrangement) system has parallels in religion, love, and music. As we saw in Chapter 6, nearly *all* religious acts involve repetitive, stereotyped motor action sequences—rituals. The distinction to be made here is that all primates engage in ritual—a way of displaying (or, in Chomsky's terminology, *conveying*) an emotional state. The ritual is the conveyance system. Religion adds to ritual a computational component—the ability to recombine, conceptualize, and recontextualize the ritual—to give it *meaning* and order. Human love is more than the ability to show attachment; it entails thinking about a hierarchy of importance, to *plan* to be in love in the future, and to communicate this plan to others.

Darwin, in *The Descent of Man*, spoke about animal "love," and it is clear from this that he is referring to conveyance and to attachment rather than to the (human) computational component:

> Animals of many kinds are social; we find even distinct species living together; for example, some American monkeys; and united flocks of rooks, jackdaws, and starlings. Man shews the same feeling in his strong love for the dog, which the dog returns with interest. Every one must have noticed how miserable horses, dogs, sheep, &c., are when separated from their companions, and what strong mutual affection the two former kinds, at least, shew on their reunion. It is curious to speculate on the feelings of a dog, who will rest peacefully for hours in a room with his master or any of the family, without the least notice being taken of him; but if left for a short time by himself, barks or howls dismally.

Later, Darwin discusses the evolutionary origin of attachment, which led to what we think of as human love:

> In order that primeval men, or the apelike progenitors of man, should become social, they must have acquired the same instinctive feelings, which impel other animals to live

> in a body; and they no doubt exhibited the same general disposition. They would have felt uneasy when separated from their comrades, for whom they would have felt some degree of love; they would have warned each other of danger, and have given mutual aid in attack or defence. All this implies some degree of sympathy, fidelity, and courage. . . . Let it be borne in mind how all-important in the never-ceasing wars of savages, fidelity and courage must be. . . . Selfish and contentious people will not cohere, and without coherence nothing can be effected. A tribe rich in the above qualities would spread and be victorious over other tribes: but in the course of time it would, judging from all past history, be in its turn overcome by some other tribe still more highly endowed. Thus the social and moral qualities would tend slowly to advance and be diffused throughout the world.

Natural selection, then, acted to select for altruism, fidelity, bonding, and those qualities that are all part and parcel of mature love. These would have been important qualities in the formation of the kinds of societies that allowed for large-scale cooperative enterprises such as agriculture, irrigation, construction projects (like grain storehouses), welfare, and courts. With the increased time it took to raise, nurture, and educate offspring, evolution had to find a way to keep the father interested in helping.

Psychologist Martie Haselton argues that love developed as a "commitment device." In one experiment, she asked people to think about how much they love their partner and then try to suppress thoughts of other people they find sexually attractive. She then had the same people think about how much they sexually desire that same partner and then try again to suppress thoughts about others. Thinking about someone you love was far more effective at suppressing thoughts of others than thinking about someone you lust after—even when it is the same person. Haselton argues that this is just what you'd expect from a neurochemical adaptation to create long-term commitment. Sex plays a role in strengthening commitment too. Most species of mammals engage in sex only during limited times, primarily when the female is fertile. Humans and bonobos are notable exceptions—we have sex even when it will not result in reproduction: during nonfertile times of the month, while the female is already pregnant, and after menopause. Zoologist Desmond Morris suggests that this was to give the male more reason to stick with one female. Add to that the shot of oxytocin that is released during orgasm, and you have a neurochemical recipe for men and women wanting to stay together.

Sex is of course a powerful drive, and implicit in Haselton's and Darwin's arguments is that sex in effect masquerades as love—whether we see love as a cultural, psychological, spiritual, or neurochemical invention, it functioned evolutionarily as a way to ensure that the product of sexual reproduction was well cared for. At the level of *society,* love has become more than just looking after one's offspring, but society itself looking after everyone's offspring—hence schools, soccer clubs, welfare, Medicare, and courts. In other words, love for one's partner and children evolved, culturally (and perhaps biologically), into the capacity to love life and fairness, goodness, and equality, and all the ideals we associate with society.

Like religion and love, music itself may be similarly decomposable into the Chomskian two-part system. Many animals produce acoustic signals that *sound* like music to us—think birds and whales, for example. But these animal "songs" almost always lack the ability to indefinitely and infinitely recombine elements according to a hierarchy, and they lack the recursion that characterizes human musics. Interestingly, the newest research shows that while animal musics *don't* have these properties so closely associated with human music, the animals *can* process sound in this way. In other words, nonhuman species contain the very foundational abilities that were for decades considered unique to human thought; it's just that they haven't (yet) learned to use them on their own. In a landmark article, Daniel Margoliash, Howard Nusbaum, and colleagues showed that European starlings can *learn* syntactic recursion. Earlier, Gary Rose found that white-crowned sparrows can assemble an entire song in proper sequence when exposed to only fragments of that song—suggesting that they possess an innate understanding of a syntactical rule for how sparrow songs are constructed. None of this is surprising given the dominant theme of evolution presented in this book—that its products tend to fall along a continuum.

Animal music is purely conveyance—its purpose is to broadcast a limited number of states-of-being. Human music comprises both conveyance and rearrangement. The computational aspect enables us to plan—to contemplate how we want to use music. We use music to convey feelings and concepts we are not necessarily feeling at the present moment. We can decide to *use* music in order to accomplish a particular goal. I can take an element from this song and combine it with that one over there. One of my favorite songs by Rodney Crowell is a song about his favorite song. In "I Walk the Line, (Revisited)," he writes about the first time he heard the famous song by Johnny Cash:

I'm back on board that '49 Ford in 1956
Long before the sun came up way out in the sticks

Note Rodney's use of internal rhymes for *board* and *Ford* (and then *before* in the next line), plus the alliteration of *back on board, forty-nine Ford,* and *nineteen fifty-six.* The phrases "back on board" and "'49 Ford" have the kind of percussive consonant quality that conveys an early rock 'n' roll feel. The second line, "Long before the sun came up," establishes the time of day, the young Crowell riding in the backseat at dawn. It also functions metaphorically, referring to this record as coming out during the dawn of rock 'n' roll. The metaphor is even sweeter to those who know that Sun Records, through its hits with Cash, Elvis Presley, Roy Orbison, and Carl Perkins, is considered by many to be the first and most important rock 'n' roll record label.

The headlight showed a two rut roadway back up in the pines
First time I heard Johnny Cash sing "I Walk the Line"

In the middle of the song, Johnny himself makes a cameo, singing the famous chorus from that famous song, his booming bass voice seemingly scraping the bottom of the range of human hearing. This sort of embedding of one phrase in another is a hallmark of human language, but turns out to have correlates in other species' vocalizations as well. Magpies and mockingbirds embed bits and pieces of other birds' songs in their own. The point is that computational modules in the brain that give us the capacity to do this fall along a continuum across species, but reach their peak in human beings. Jazz players quote other music all the time, a trick they borrowed from the great composers, including Haydn and Mozart, who embedded pieces of other songs into their own. And to my ears, nothing sounds as sweet as Rodney's music—and wordplay in this song, a moving and loving tribute to memory, to music, and to human creativity.

Human music has hierarchical structure and complex syntax, and we compose within that constraint. Music, like language and religion, contains elements shared with other species and also elements unique to humans. Only humans compose a song for a particular purpose, made up of elements found in other songs. Only humans have the vast repertoire of songs (the average American can readily identify more than one thousand different songs). Only humans have a cultural history of songs that fall within six distinct forms.

It is important, when considering animal music, to distinguish beween musical expression and musical experience. In other words, many animals express themselves in ways that sound musical to us, but are really functioning as conveyance and messages among themselves; there is no evidence that they *experience* music as the aesthetic or creative artform that we do.

It is advisable also to consider that *music* per se is not what evolved, but rather, music comprises components each of which has undergone a specific and probably separate evolutionary trajectory. Pitch, rhythm, and timbre are processed in separate parts of the brain. They come together later during processing, and melody, a higher-order concept, is constructed from these, influenced by changes in any one, or any combination, of the lower-level features. Music as we know it emerged in evolution after these component processes were already in place.

Evolution endowed the musical brain with a perception-production link that most mammals lack. This motor-action-imitation system gives us the ability to take something in one sensory domain and figure out how to create it with another. We *hear* music, then *sing* it. Every song you know how to sing, every word you speak, you reproduce with your own voice based on something you originally heard. Humans (and only a few other species, such as parrots) are able to turn what we hear into what we reproduce vocally. In spite of the high intelligence of mammals, most do not have the capacity to imitate a sound they've heard (exceptions include humpback whales, walruses, and sea lions). This vocal learning ability is believed to have come from an evolutionary modification to the basal ganglia, causing a direct pathway between auditory input and motor output.

What's amazing is that, at the beginning of the twenty-first century, we can see for the first time the emergence of human culture reflected in the genome. The classic method of examining fossilized skulls continues to reveal new surprises as well. For example, fossil evidence indicates Brodmann area 44 (BA44)—a part of the frontal cortex that is important for auditory motor imitation via mirror neurons there—may well have been in place 2 million years ago, long before *Homo sapiens* (who didn't emerge until about 200,000 years ago). In other words, the neural mechanisms for language were in place long before they were fully exploited. The FOXP2 gene, closely associated with human language, existed in Neanderthals; a form of it is also found in songbirds. A genetic variant in microcephalin (a part of the genome that encodes for brain development) has been identified. It emerged approximately 37,000 years ago—what we think of as the beginning of culturally modern humans. This coincides, not coincidentally, with the appearance in the archaeological record of artistic artifacts and bone flutes. A second variant of microcephalin arose around 5,800 years ago, corresponding with the first record of written language, the spread of agriculture, and the development of cities. Who knows what we will uncover about this story in the coming decade, but we now know that FOXP2, BA44, and microcephalin variants are part of the evolutionary changes that primed the creation of the musical brain.

The human motor/action imitation system also evolved the ability to imitate with delay—in the absence of the original model. This underlies language, music, religion, painting, and the other arts. Only humans represent, symbolize, and signify things that are not there. (As with recursion, some animals can be trained to do this, indicating that the neural structures are in place, but have not yet been exploited.) With this follows the inevitable question: If I can think about things that are not here—my loved one off on an expedition, that tiger that attacked my friend—are there things not here that I *haven't* thought about? Are there other worlds, are there entities, that exist outside my experience? This led, as we saw in Chapter 6, to a yearning for spiritual knowledge. It also leads to a yearning to form long-term bonds with loved ones, to gain assurances that they'll stay with us and come back to us.

Human consciousness, a product of *representation* and thus another feature of the musical brain, appears to be different from animal consciousness as well. As we saw in Chapter 6, animals live in a continuously unfolding present, with no ability (as far as we know) to reflect on the past or plan for the future. Some have argued that everything we humans do is done unconsciously, and that one of the roles of consciousness is to spin a story after the fact about what we did and why we did it. Patients whose two cerebral hemispheres have been surgically separated (for the treatment of intractable epilepsy) often exhibit this kind of post facto rationalization, supporting this notion, and the rationalization typically occurs when the right hemisphere causes the person to do something and the left hemisphere (the hemisphere with language) is left holding the explanatory bag.

Many neuroscientists have been looking for the seat of consciousness in the brain, and I believe that they will never find it—not because consciousness doesn't exist, but because it is not localizable. Just as we don't expect to find "gravity" at a particular location in the middle of the earth, we shouldn't expect to find consciousness at a particular place in the head. I take as an initial assumption the view expressed by my colleague, the McGill University physicist and philosopher Mario Bunge and by other contemporary philosophers, including Paul and Patricia Churchland and Daniel Dennett, that there are no immaterial, vitalistic, or supernatural processes involved in creating the experience we call consciousness; rather, it is a process that arises from the normal functioning of neurons in the human brain.

When biological complexity arises from simpler forms in small steps, we call it evolution. When a wholly unexpected property— such as human consciousness—arises from a complex system, we call it emergence. Ant colonies exhibit an emergent intelligence, finding food, disposing of waste, feeding their queen—yet no ant can be said to "know" in any meaningful sense what it is doing or what the colony is doing. In this respect, ants can be thought of analogously to neurons in the human brain. There is no neuron in your brain that knows your name, and no other neuron that is even bothered by this fact. There is no neuron that knows how old you are, where you were born, what your favorite ice cream is, or whether you are too hot or too cold right now. Neurons just don't work that way. Hundreds of thousands or millions of neurons need to come together to encapsulate, store, or provide information.

An individual ant, like an individual neuron, is just about as dumb as can be. Connect enough of them together properly, though, and voilà! The system-as-a-whole demonstrates spontaneous intelligence. From the firings and interconnections of billions of neurons, we can look at life and our place in it, we can even think about the nature of our own thoughts. Thoughts emerge from brains, but the process remains mysterious. Emergence has even been invoked as the source of life itself. Consider the so-called primordial soup millions of years ago—a bubbling, boiling compost of carbon, nitrogen, oxygen, hydrogen, proteins, and nucleic acids. The first single-cell organisms arose here, and biologists now believe that *life* arose as an emergent property of the complexity of those first molecular compounds.

Most scientists and philosophers agree that human consciousness is qualitatively different from animal consciousness—we have a unique type of self-consciousness, and an ability to contemplate our own existence. Consciousness is not a thing to be separated out from the normal workings of our brain, the comings and goings of thoughts, perceptions, and mental states. As essayist Adam Gopnik says, "consciousness is not the ghost in the machine; it is the hum of the machinery."

Some of our appreciation of music is not conscious, but rather, forms part of our subconscious awareness of the world. For example, as David Huron and Ian Cross argue, human music functions as an *honest signal,* as we saw in Chapter 5. This is a concept from biology that relates to the extent to which a communicative signal from one organism to another can be faked. There might be reasons for an organism to convey untruthful information—this is essentially what chameleons do when they change color to blend in with the background to avoid predators, or what possums do when they play dead. Primates may wish to deceive one another for a number of reasons, including hoarding food or trying to rise in a complex social hierarchy. If music is an honest signal, a person who is singing would be less able to fake emotions. Put another way, we *believe* a message that comes to us through song more than one that comes through speech. For reasons that aren't fully understood, we seem to be exquisitely sensitive to the emotional state of singers. This may be related to unconscious cues of stress that are conveyed through the vocal cords when a singer is lying, but this is a topic that requires further investigation.

The honest signal hypothesis is particularly relevant for love, and may explain why love songs move us so. When someone tells us they love us, we may have our doubts. When they sing it, all our doubts seem to melt away. This may be an evolutionary and biological inheritance, something that is beyond our powers of rational control or conscious influence—singing matters. This could account for why people become *furious* when they find that singers are lip-syncing. It may also explain our fascination with the private lives of rock singers (much more so than other band members): If they are not living the life they profess in their songs, our truth detectors go wild.

The connection between truth and love is obvious—when we love, we make ourselves vulnerable (and there are many songs that reflect this naked emotional state). Real love requires an almost irrational trust and faith in another person. We never *really* know if our partner has been faithful, has been stealing money from us, or has an ulterior motive. As Will Smith's character says in the movie *Hitch,* love is like jumping off a cliff. These concerns take time, sometimes years, to assuage, and relationships would never make it that long if we didn't suspend suspicion temporarily and take the risk of letting the other person into our lives and hearts. Different people resort to different mechanisms, some psychological, some practical, to bridge this gulf of trust. Some refuse to make themselves completely vulnerable to their partner, exchanging safety for lack of intimacy. Some—who are otherwise very well protected in all other dealings, in business and friendships—throw caution to the wind over and over in their love lives. Each new love is a new beginning; "the only love worth having is one that is complete and total," we say to ourselves. Others insist on prenuptial agreements. "I've never seen a prenup I couldn't break," a lawyer once told me, "so why bother in the first place?"

Love requires that we give our partner the benefit of the doubt. Lipstick on the collar or missing condoms are a warning sign, but most "signs" are more ambiguous: staying late at the office, a partner who doesn't answer his hotel room phone at midnight. (Did the hotel ring the wrong room? Did he turn the phone off? Is he in the shower? Is he with another woman?) Every month dozens of signs *could* point to our lover's dishonesty, infidelity, or even simpler violations of the relationship contract such as not revealing her true income or true feelings. Love, it has been said, requires a certain amount of self-delusion. Animals don't have this problem because the animal brain doesn't ruminate, try to fit pieces together like a detective, try to figure out is she right for me? The animal brain bonds on a combination of instinct and pheromones. The role of instinct and pheromones in human mating decisions is also very strong, but it seems to hang in an uneasy balance with rationality, or at least self-delusion and justification masquerading as rationality. This is one of the reasons that the song "I'll Get You Back" by Juliana Raye (brilliantly produced by former ELO frontman Jeff Lynne) is so bitingly ironic and funny:

When you ran away from me you never looked to see
I was right behind you running just as speedily
Slow down, would you tell me where you're going
'Cause I need to know if you'll be back in time for supper
I cooked your favorite

The vocal is delivered with a kind of upbeat, twisted, clueless delusion. Her boyfriend is running out of the house—not walking but *running*—and she chases after him to find out when he's coming home for dinner. And oh, by the way, she yells after him, "I COOKED YOUR FAVORITE!"

In the second verse she tells us that she knows about all the affairs he has had, but she doesn't care, as long as he practices safe sex. Launching into the singsongy, nursery-rhyme chorus, she sings:

I will get you, I will get you, I will get you back
I'll get you back, I'll get you back, I'll get you back
I'll get you back!

As the last note of "back" reaches a pitched crescendo, the highest note of the song, the evil quiver in her voice forces us to confront the ambiguity in the chorus: Does she mean that she is going to "win him back" or that she is going to "get back at him" by some awful retaliation? The second meaning is invoked more strongly as we learn of a heart-stopping violation of trust right along with the singer in the last verse:

Sister Mary always had the kindest words to say
She said when she looked at you her doubts would melt away

I swear, Sister Mary's baby looks a lot like you, you know
Oh! Say, it isn't so!

I will get you, I will get you, I will get you back
I'll get you back, I'll get you back, I'll get you back
I'll get you back!

But something in the delivery tells us that she still wants him, that she still hopes somehow to win back this man who has mistreated her so. That all will be forgiven in the name of love. She got it bad, and that ain't good. Love does require this kind of devotion, even blind commitment, although typically not such self-delusion. The Australian band Mental as Anything played with this same theme in their lyric "If you leave me, can I come too?"

There are love songs to reflect four stages of love: I want you, I got you, I miss you, and it's-over-and-I'm-heartbroken. Love songs reflect the different kinds of love as well: the Romeo-and-Juliet love (I'd kill myself for this person); the more mature love of being together for decades and looking back; and the love of ideals, such as of country. The dominant theme of popular music for the last fifty years seems to be love in all these forms. Songs written by Cole Porter, Irving Berlin, Lennon and McCartney, Dylan, Mitchell, Cohen, Wainwright. And the songs performed by Diana Ross, the Temptations, the Four Tops, Frank Sinatra, Ella Fitzgerald, Mariah Carey . . . Some love songs celebrate the earliest stages of romantic love:

Oh yeah I'll tell you something, I think you'll understand
When I say that something, I want to hold your hand
I want to hold your hand, I want to hold your hand

and they capture the giddy, lighter-than-air feeling of first loves and crushes:

And when I touch you I feel happy inside
It's such a feeling that my love, I can't hide, I can't hide, I can't hide!

Songs reflect the fear we have of becoming vulnerable to another person, and to love's capriciousness—an implicit acknowledgment that the wonderful feelings can apparently end at any time and without warning. Some songs express pure denial (such as "I'm Not in Love" by 10cc):

I'm not in love, so don't forget it
It's just a silly phase I'm going through
And just because I call you up
Don't get me wrong, don't think you've got it made
I'm not in love, no-no

I keep your picture upon the wall
It hides a nasty stain that's lyin' there
So don't you ask me to give it back
I know you know it doesn't mean that much to me
I'm not in love, no-no

On the opposite side of things, this vulnerability is sometimes intentionally cultivated, David Byrne notes. Rather than denying our feelings and vulnerability, we sometimes say to our love interest, "I am yours" (Stephen Stills in "Suite: Judy Blue Eyes") or "Take all of me . . . I'm no good without you" (Marks and Simons), or as Irving Berlin writes in "Be Careful, It's My Heart," from the musical *Holiday Inn*:

It's not my watch you're holding, it's my heart.
It's not the note I sent you that you quickly burned.
It's not the book I lent you that you never returned.

As sung by Frank Sinatra, Rosemary Clooney, Bing Crosby, and others, the song reminds us of the helpless condition love can put us in, underscored by the repetition of the song's title, "Be Careful, It's My Heart," as the lyric continues:

The heart with which so willingly I part.
It's yours to take to keep or break,
But please, before you start, be careful, it's my heart.

"But what really makes that song," David says, "are the harmonies. Love song lyrics can be the corniest thing in the world. The harmonies add tension that keeps the lyrics from seeming too corny. Too often, love song lyrics can't stand on their own two feet apart from the melody and the harmonies. Another great love song is Joni Mitchell's 'A Case of You.' Those lyrics are great—they tell a whole little story, a mini novel inside a three-minute song."

"I think some the greatest love songs of the twentieth century were those written by Hugo Wolf," Stanford composer Jonathan Berger says. Wolf wrote hundreds of songs at the turn of the twentieth century. They are so complex harmonically as to be virtually unsingable, but they raise the interplay of melody, harmony, and lyrics to a very high level.

"Tonality is dissolving as he is writing," Berger continues, "and he's at that cusp where everything he writes is tonal, but he doesn't use tonality in terms of 'Let's get to the dominant and get over with it.' Instead, he uses tonality in a very symbolic way. For example, he has a song called 'The Forsaken Maiden.' She's waking up in the morning and *in the song* she starts singing this love song. And, over the course of the song, realizes that she's been forgotten, and forsaken, and left. So it's this process of dream to awakeness, of *having* the love and *losing* the love, and the melody and the harmony are constantly playing with this 'Am I dreaming, am I awake? How aware am I?' The music defines everything in the text. In other words, the song itself, the way it is composed, becomes representative of love itself."

Many of popular music's most memorable and emotional songs deal with the sexual, lustful side of love. "As soon as music first emerges in cavemen and it has rhythm," Rodney Crowell says, "you have the sense of *sex* in it, because what is the most obvious human activity that has a rhythmic component? One of the earliest songs must have been a song about sex." One of the oldest songs that we know of is in fact one with a sexually charged lyric; dating back six thousand years, it is part of the extraordinarily graphic poems and songs written by Inanna, queen of Sumeria, about her beloved Dumuzi.

In the forties and fifties, artists recorded risqué R&B records such as Bull Moose Jackson with "My Big Ten Inch" (about a phonograph record of a band that plays the blues) and Dinah Washington singing "Big Long Slidin' Thing" (ostensibly about a trombone player). Early heavy metal songs made no attempt at clever innuendo or subtlety: "Whole Lotta Love" and "The Lemon Song" (Led Zeppelin), "Hot Blooded" (Foreigner), "Feel Like Makin' Love" (Bad Company). The Stranglers sang of "walking on the beaches/looking at the peaches," and Joni Mitchell about how a lover "picked up my scent on his fingers/while he's watching the waitress' legs." More recently, the Magnetic Fields released the ambitious and quirky *69 Love Songs* cycle; many of them might better be described as lust songs, such as "Underwear": "A pretty boy dressed in his underwear/if there's a better reason to jump for joy/Who cares?" Lusty songs are not limited to American or even Western culture, from the Pakistani singer Nadia Ali Mujra's *"Chuupon Gye Chuupon Gye"* (I want to suck on the mango) to several songs on the Cantonese Top 40.

"I think the most lecherous song in our lifetimes is 'Sweet Little Sixteen' by Chuck Berry," Rodney says. "You've got this adult lusting after a sixteen-year-old, almost a predator. And the song exists as a pivot tune between innocent rock and roll—like 'Rock Around the Clock'—and a kind of perverted sexual lasciviousness. The lyrics themselves are innocent of course, but every eighteen-year-old boy and sixteen- or seventeen-year-old girl knows exactly what this dirty old man is talking about. 'All the cats wanna *dance* with Sweet Little Sixteen' is a metaphor of course. And we now know enough about the man who is Chuck Berry—I mean, the guy who later put a hidden camera in his bathroom is the guy who wrote 'Sweet Little Sixteen.' And of course all the chords have that flat seventh, which is diabolical sounding."

Many people report that music triggers memories long ago buried, and this seems especially true of popular love songs. This may be a modern phenomenon, and from what we now know about the neurobiology of memory, it is a product of the way that human memory works. The prevailing view among memory theorists is that nearly every experience we've had is encoded in memory; the trick is getting it *out*. What you want in a retrieval cue—jargon for the thing that helps you pull a memory up from your brain—is something uniquely associated with a time or place or event.

Pop songs, because they are typically played over and over again on the radio for a short period of time, make ideal cues for this, as do smells. A song like the national anthem or "Happy Birthday" can certain trigger memories, but a song you haven't heard since you were fourteen years old is more likely to trigger deep, buried memories.

"In that first moment when we're sexually attracted or romantically attracted or both, whatever song is there in the background takes on an absolutely important meaning and that meaning will never ever vanish or fade," Jonathan Berger says. "And I think what that is saying, is that there is always a long game. There's always an 'I didn't do it, I didn't get it' that persists all the way through life. I have this strong memory associated with Freda Payne's 'Band of Gold.'

"I was in high school and I was working that summer; my best friend and I had a job of delivering things to construction sites. I used to hide the brooms and shears behind the closet and jump into the car with him and we'd go off and look at girls and cruise. And that was the song. There is this absolutely sheer emotional teenage attachment to that song that's never ever gone. The recording just drips with energy. She distorts the mic and it's that beautiful warm analog distortion. And you can't really tell what all the instruments are, they're so distorted. And you just feel like you're in the Apollo or something. So it's an exciting track. And it revolves around love lost and broken love promises—all she's left with is the ring he got her, the 'band of gold.'"

I posed the question earlier of why evolution hit upon sound in general and music in particular to help us communicate deep emotions such as love. The answer lies in music's ability to serve as an honest signal, to hold memories, and to move us neurochemically. Consider the question from another angle: What if evolution could discover a way that you could stick in your loved one's thoughts even when you're not around, creating a strong emotional tie in your absence that includes just the right mix of neurochemicals to promote feelings of comfort, fidelity, and trust, and raise your loved one's mood all at the same time? Evolution would have to draw on existing structures through a series of minute, tiny adaptations. It would want to tap into a primitive motivational system if it could. It should play on a balance between fears of abandonment and comfort in togetherness, between love and lust, and it should stimulate the higher cognitive system with a sense of play, order and reorder, figure and reconfiguration. Music does all of this for us, and love songs imprint themselves in our brains like no others. They speak of our greatest human aspirations and loftiest qualities. They speak of setting aside our own ego and desires in the service of something great—of caring about someone or something more than we care about ourselves. Without the innate ability to have such thoughts, societies could not be built.

Each of us is the end product of thousands of years of a genetic arms race to determine which genes will survive. Genetic real estate is expensive—only those genes that help us in some way are likely to keep their chromosomal homes, otherwise they are crowded out.

Each of us is selected for finding babies cute, for being able to attract a mate and avoid predators, for finding fresh fruit tasty, for appreciating artistic representations of the world, and so on. Sex doesn't intrinsically feel good—those of our ancestors who enjoyed it are those whose children lived to tell about it. We're also selected to find abstract thought and imaginative thinking to be enticing, appealing. Recall Dawkins's mantra: No one alive today has a single ancestor in his or her past who died in infancy. We are the champions, my friend!

Love in all its various forms is ultimately about caring—caring so deeply about another person, group, idea, or place that we would be willing to sacrifice our own health, comfort, and even life for it. One hallmark of great art is the amount of care that we sense has gone into it. When people scoff at modern painting, their typical objection is that it looks as though the painter simply threw paint on the canvas *with no care.* We find ourselves drawn to art that looks as though the artist struggled with it, put a great deal of thought into it—*cared* about it. In visual art this can be because of the obvious time it takes to create an impressionistic or photorealistic painting. One of the most engaging art installations I've ever seen was by the artist Michael C. McMillen, who created, from tens of thousands of pieces, a 1950s suburban garage/workshop inside the Los Angeles County Museum of Art. The detail and care put into it is what was overwhelmingly impressive to me and viewers I stood alongside that day.

In music, many of us respond strongly to the caring of the musical artist for the medium of music—through the textural layerings of Pink Floyd or the Beatles, or the sheer technical skill of instrumentalists such as Keith Jarrett and Cannonball Adderley. In this book, I've endeavored to include examples from music all over the world. Although my main argument is that there are six kinds of songs that have shaped the history of human civilization, I am not so reductionist as to say that there are only six individual songs that matter; rather, I believe that the breadth of music and human musical expression is what is most impressive. Although the important functions of music can be described in these six categories, the specific ways that people from different musical cultures have found to make music are very diverse.

I co-taught a course at Stanford in 1991 called "Technology and Musical Aesthetics" with Bob Adams, a mechanical engineer friend. The course reviewed the history of musical instrument design from ancient Greek water-driven pipe organs to the newest digital synthesizers. In considering those technological advances that had had the greatest impact on music, Bob and I agreed on two as far and away more powerful and influential than any of the others. It wasn't the discovery of equal temperament or of amplification, not the technology to make keyboards. First was music notation, which allowed for musical works to be preserved, shared, and remembered. The ancient Greeks had the oldest system we know of, and modern notation (more accurate and clearer) traces its development to five to eight hundred years ago.

Second was the innovation of recording, first on wax cylinders (the Edison phonograph), then on acetate and vinyl, later on tape, and now on CDs and digital files. But regardless of the format, I believe that recording was truly one of the most influential developments in the entire history of music making. The reason is that the act of recording fundamentally transformed the way that people thought about performances. A song could be performed only once, and that exact performance could be played back over and over again, theoretically an infinite number of times, and all around the world. The recording could be played in places the performer wasn't, and even after the performer was no longer alive. More importantly, the recording introduced the idea of a "master performance," a single canonical, iconic version of a song. Although groups like Phish, Dave Matthews Band, and the Grateful Dead made careers out of performing songs differently each time they play (something that is de rigueur in jazz), the standard way of thinking about popular music for the past fifty to a hundred years has been that a single "official" version of the song exists. This is the version that entire communities learn, and it results in a collective sharing of music on a truly unprecedented, global scale. And recordings have become their own distinct aesthetic objects, with a sound and auditory sensuality all their own. I've had a lifelong love affair with recordings, and I see them as an art form in themselves.

In describing the world in six songs, I've intentionally avoided becoming distracted by questions such as "What are the greatest/most popular songs of all time?" or even—to me as a musician—the more interesting question of "What are the most *influential* songs of all time?" (They aren't necessarily the same question of course.) The most influential song of all time, by definition, I think, would have been the first song that one of our ancestors heard another person sing or play, and that got stuck in his head, making him want to sing it back again through the perception-production system that evolution had endowed him with. In our own lifetime, the blues from the early 1900s—whatever the first blues song was is lost to history—may earn the prize, having spawned or at least influenced jazz, R&B, gospel, rock, metal, bluegrass, and country. Schoenberg's *Pierrot Lunaire,* often regarded as among the first atonal pieces (1912) had an important impact on classical music, which in turn influenced later twentieth-century jazz, rock, and experimental music. But here, I've been more interested in the broad, long arc of musical history, going back tens of thousands of years, than the workings of the past century, although whenever I could I've tried to choose examples from the last century to illustrate points, in order for us to have some common ground.

My musical tastes were shaped by a childhood spent growing up in the Northern California of the 1960s. In my own idiosyncratic view, three contemporary, Western musical artists who represent the very peak of love songs are Alex de Grassi, Guy Clark, and Mike Scott.

There are certain musical events that changed the way I would hear for the rest of my life. The first time I heard the Beatles, the first time I heard a Cannonball Adderley solo, and the first time I heard the acoustic guitarist Alex de Grassi. Tom Wheeler wrote in *Guitar Player* magazine, the bible for serious guitarists, that Alex's technique is "the kind that shoves fellow pickers to the cliff of decision: should I practice like a madman or chuck it altogether?"

Alex's compositions and playing ability are extraordinary from a technical standpoint, but more importantly, they move me to tears of joy and sadness, and sometimes I can't tell the difference. As "Another Shore" (from *The Water Garden*) plays, happiness, melancholy, tranquillity, exhilaration, awe, and appreciation fill me all at once. His compositions reinforce a conviction that music really is, as Tori Amos says, the voice of the universe. And it sometimes humbles me in seeming to be a voice that I can understand but not speak.

Often, listening to Alex's records, I am overwhelmed by the realization that I know nothing about music at all. If I somehow get pulled out of my listener's reverie and reclaim some of my self-awareness, something inside me says I will never be that good! I cannot imagine what I would need to do to bridge the gulf between Alex's writing and mine; between his guitar playing and mine; in general, between his musicianship and mine. It's an odd feeling because I think that I can imagine how the gulf between me and, say, Tom Petty or Neil Young could be bridged; I know I'll never be as good as they are either, but the chasm at least seems bridgeable with a reasonable and understandable course. Paul Simon told me that a truly great artist in any medium is one whose influences and ideas are opaque to the audience, someone who seems to have tapped into something entirely different from the rest of us. And that is Alex. Alex cares deeply for music and for the guitar, and spends thousands of hours perfecting his composing and playing. The love shows.

I think of Guy Clark as a craftsman. What he does with melody and with lyrics are something to marvel at, because he does it with such *economy.* If you look at fine furniture, it's not always the piece with the fancy carvings that you like, it's something that's just elegant because of the lines of it, the understated beauty, and that's how I feel about "The Randall Knife"—and indeed his songs in general. Like a fine carpenter, Guy takes great care in constructing his songs, going over and over them until they're *just right.* You can tell that he loves the process, that he loves songs, and this is reflected in the evident care he takes in all aspects of the song: the writing, the performance, even the recording.

"The Randall Knife" is a story about a father and a son. The father has died, and the son is talking about the death of the father, and in a sense the song symbolizes both the *death* of the relationship and the *birth* of the relationship. Obviously, there's the *physical* death of the relationship as the father is no longer of this world, but you hear in the song a real transformation in the narrator's perspective that leads you to believe that the relationship with his father is now only *beginning.*

My father had a Randall knife
My mother gave it to him
When he went off to World War II
To save us all from ruin
If you've ever held a Randall knife
Then you know my father well
If a better blade was ever made
It was probably forged in hell

My father was a good man
A lawyer by his trade
And only once did I ever see
Him misuse the blade
It almost cut his thumb off
When he took it for a tool
The knife was made for darker things
And you could not bend the rules

He let me take it camping once
On a Boy Scout jamboree
And I broke a half an inch off
Trying to stick it in a tree
I hid it from him for a while
But the knife and he were one
He put it in his bottom drawer
Without a hard word one

There it slept and there it stayed
For twenty some odd years
Sort of like Excalibur
Except waiting for a tear

My father died when I was forty
And I couldn't find a way to cry
Not because I didn't love him
Not because he didn't try
I'd cried for every lesser thing
Whiskey, pain and beauty
But he deserved a better tear
And I was not quite ready

So we took his ashes out to sea
And poured 'em off the stern
And threw the roses in the wake
Of everything we'd learned
When we got back to the house
They asked me what I wanted
Not the lawbooks, not the watch
I need the thing he's haunted

My hand burned for the Randall knife
There in the bottom drawer
And I found a tear for my father's life
And all that it stood for

We get a sense of who the characters are with very few words. I've heard this song so many times, but it moves me to tears every time. The love he had for his father, the love he has for the medium of song, the love for those things that make life so difficult and yet so precious.

"Bring 'Em All In" by Mike Scott (the lead songwriter and singer for the Irish band the Waterboys) is to my ears among the most perfect love songs ever written. It is the yearning of one human to feel at one with the world, to embrace all that is contained in it. It is a love song to all of us, to the good and the bad, to the great and the small. It is the song of one man alone with his thoughts, by himself, trying to reach out to become connected. Out of his loneliness and despair he discovers the deepest love, the love of the idea of living, the love of love itself, a willingness to open his heart to *everything*, even the pain that he knows will be let in when he does so.

The song opens with a rapid, almost Spanish strum (technically very difficult to do), played with the fingers and not a pick, to give a delicate gentleness to the moment the fingers strike the strings.

Chorus:

Bring 'em all in, bring 'em all in, bring 'em all in
Bring 'em all, bring 'em all in to my heart (2×)

Bring the little fishes, bring the sharks
Bring 'em from the brightness, bring 'em from the dark

(Chorus)

"Bring 'em all in" functions as a mantra. The melody scoops up on the word "heart," as though lifting up objects from the floor or from the depths of the sea. The half-whispered voice sounds like a prayer, an intimate song to oneself and one's creator.

Bring 'em from the caverns, bring 'em from the heights
Bring 'em from the shadows, stand 'em in the light

(Chorus)

Bring 'em out of purdah, bring 'em out of store
Bring 'em out of hiding, lay them at my door

(Chorus)

In a double chorus in the middle, the singer's voice becomes even softer, nearly crying, the song takes on the quality of a lamentation, the pleas of a man who is spiritually dying and opening his heart for one last hopeful time as he sings the last verse:

Bring the unforgiven, bring the unredeemed
Bring the lost and nameless, let 'em all be seen
Bring 'em out of exile, bring 'em out of sleep
Bring 'em to the portal, lay them at my feet

It is the love of our existence that is the highest love of all, the love of humanity with all our flaws, all our destructiveness, all our petty fears, gossip, and rivalries. A love of the goodness that we sometimes show under the most difficult stresses, of the heroism of doing the right thing even when no one can see us doing it, of being honest when there is nothing to gain by it, of loving those whom others might find unlovable. It is all this, and our capacity to write about it—to celebrate it in song—that makes us human.

Notes

CHAPTER 1

p. 3 **"Americans spend more money on music than they do on prescription drugs or sex . . ."**
Huron, D. (2001). Is music an evolutionary adaptation? *Annals of the New York Academy of Sciences* 930: 51.
". . . the average American hears more than five hours of music per day."
The average American watches five hours of television per day, and that alone has to account for a lot of music listening—comedies, dramas, commercials, even news programs are presenting music nonstop. Add in music in public places such as train stations, restaurants, office buildings, and parks, and music wafting in from a neighbor's yard or apartment, and you've got lots of music. See acnielsen.com for statistics.

p. 5 **". . . a catchy song about a murderous psychopath who kills the judge at his own trial . . ."**
Lennon, J., and McCartney, P. (1969). Maxwell's silver hammer [Recorded by the Beatles]. On *Abbey Road* [LP]. London: Apple Records.
". . . a song promising to keep a promise . . ."

Prosen, S. (1952). Till I waltz again with you [Recorded by Theresa Brewer]. On *Till I Waltz Again With You* [45rpm record]. Coral Records.

". . . a song mourning the loss of a parent . . ."

Crowell, R. (2001). I know love is all I need. On *The Houston Kid* [CD]. Sugar Hill Records.

p. 7 **"Ten thousand years ago humans plus their pets and livestock accounted for about 0.1 percent of the terrestrial vertebrate biomass inhabiting the earth; now we account for 98 percent."**

MacCready, P. (2004). The case for battery electric vehicles. In *The Hydrogen Energy Transition: Cutting Carbon from Transportation*, edited by D. Sperling and J. Cannon. San Diego, CA: Academic Press, pp. 227–233.

I first heard about this from Daniel Dennett, at a talk he gave at McGill University in 2006.

"We're also a highly *variable* species."

Even within the contemporary United States, there are subcultures of people who eat dirt, avoid eating anything with a face, or who eat only food that has been ceremonially blessed by their religious leaders. Americans speak more than three hundred different languages; some read right to left, some left to right, some up to down, many not at all. And music? A survey I conducted of one thousand Canadian college undergraduates last year revealed that they identified sixty separate genres of music overall that they listen to, genres that were distinct and ranged from ancient Sufi music to Swedish death metal; from the indigenous folk singing of the Ural Mountains to the heavily processed vocal stylings of Nine Inch Nails. If this much musical variety emerges from North American college students, imagine the variety that exists worldwide and age-wide.

p. 8 **"By definition, a 'song' is a musical composition intended or adapted for singing."**

This book is called *The World in Six Songs,* not *The World in Six Musical Works*. Musicologists generally make a distinction between "song" and longer musical forms, and a "song" is typically understood to have words. This distinction is meant to imply that *song* is a subset of *music,* and this does seem to follow intuition. Most of us don't think of Wagner's *Ring Cycle* or Beethoven's Fifth Symphony as *songs*. But suppose you had never heard the latter before, and then one day, you hear your grandfather puttering around the yard humming "bum-bum-bum-baaah, bum-bum- bum-baaah." You would be perfectly justified to ask him just what is this *song* you hear him bum-bum-bumming, and neither the language police nor the musicology marshals should have any cause for alarm that important communicative barriers, essential to the maintenance of an orderly society, have been breached. Just as the Aleutians are reputed to have twenty words for snow, so do we have many words for different forms of music: jingles, ditties, tunes, melodies, airs, anthems, arias, odes, ballads, canticles, carols, chants, chorales, choruses, hymns, lullabies, numbers, operas, pieces, rhapsodies, refrains, cantatas, shanties, strains, verses (not to mention words for specialized musical forms such as sonatas, cantatas, symphonies, quartets, opuses) . . . and the distinctions between them all can be interesting. These different musical forms typically convey different types of messages—we don't expect an anthem to lull a baby to sleep, nor do we expect a chorale to sell tickets to a Monster Truck smashing competition. But what is the use of this distinction between *song* and *music*?

Actually, the Aleutians don't have any more words for snow than we do in English. See the excellent and amusing essay: Pullam, G. (1991). The great Eskimo vocabulary hoax. In *The Great Eskimo Vocabulary Hoax and Other Irreverent Essays on the Study of Language,* edited by G. Pullam. Chicago: University of Chicago Press, pp. 159–174.

". . . John Denver took Tchaikovsky's Fifth Symphony and added lyrics to the melody?"

The melody for "Annie's Song" is very close to the main theme for Tchaikovsky's Fifth Symphony, *Andante cantabile,* during the part when John Denver sings "You fill up my senses." For the next line, "like a light in the forest," Denver writes a variation of that theme, staying well within Tchaikovsky's original harmonic context. Denver, J. (1974). Annie's song. On *Annie's Song* [CD]. Delta Records. (1997).

Tchaikovsky, P. I. (1888). Symphony no. 5 in E minor, op. 64 [Recorded by M. Jansons (Conductor) and the Oslo Philharmonic Orchestra]. On *Tchaikovsky: Symphony No. 5* [CD]. Chandos Records. (1992).

p. 9 **". . . 'Happy Birthday' has been translated into nearly every language on earth (even into Klingon, as viewers of *Star Trek: The Next Generation* can attest; the song is called *'qoSlIj DatIvja'* ".**
For a pronunciation guide, see http://www.kli.org/tlh/sounds.html.

p. 12 **"I have actually known people named Maggie Mae, Roxanne, Chuck E., and John-Jacob (think Rod Stewart, the Police, Rickie Lee Jones, and an old children's song), and they are astonished when people sing these songs to them as though no one has ever thought to do this before."**
In a fit of what I thought was my own cleverness, I asked Rosanne Cash on first meeting her if it was true that she had a brother named "Sue." I was making reference to the song written by Shel Silverstein and made famous by her father, Johnny Cash, "A Boy Named Sue." Rosanne rolled her eyes, and lest the gesture was lost on me—after all, at this point, she had every reason to think that I was dim-witted—she said, "You have no idea how many people have asked me that." I suppose I am dim, because it took me several months of playing this encounter over in my mind (and wishing I had said something more intelligent to so beautiful and charming a woman) that I realized that the joke, weak as it was, was poorly formed. The hero of the song is a man who *himself* was named Sue by his father. Johnny is singing the song *as* Sue. In fact, at the end of the song the narrator sings "If I ever have a son I think I'll name him Bill, or George—anything but Sue!" So the proper way to make the joke would have been to ask Rosanne if it was true that her *father* was named Sue, not her brother. When I next met with her—thank goodness, she either has a huge heart full of forgiveness or forgot me between meetings—I pointed this out to her and she said that she had already been through this dizzying chain of reasoning, and that it continued to surprise and amaze her that virtually every one of her tormentors asked her about a brother, no one getting the (fictional) familial connection right.
Silverstein, S. (1969). A boy named Sue [Recorded by Johnny Cash]. On *This Is Johnny Cash* [LP]. Harmony Records.

p. 13 **"When a man loves a woman / Spend his very last dime / . . ."**
Lewis, C., and Wright, A. (1966). When a man loves a woman [Recorded by Percy Sledge]. On *When a Man Loves a Woman* [LP]. Muscle Shoals, AL: Atlantic.

p. 15 **"It is unlikely that either language or music was invented by a single innovator or at a single place and time . . ."**

Individual songs are "invented" of course, as are individual words such as "Vietnamization," or "soundscape," but this is not the same thing as inventing music or language themselves.

". . . perspective-taking . . . "

Psychologists call this "theory of mind," a term introduced by David Premack and Guy Woodruff in 1978 and used commonly in the developmental psychology literature. It is similar to what Piaget called "objectivity." I find "theory of mind" to be a bit too jargony, so I'm going to use the term "perspective-taking." Stated this way, the link becomes clearer to Einstein's recognition of the importance of the perspective of the observer; this is fundamentally the same notion, that different observers will perceive the same events differently.

Premack, D. G., and Woodruff, G. (1978). Does the chimpanzee have a theory of mind? *Behavioral and Brain Sciences* 1(4): 515–526.

p. 16 ***". . . octave equivalence . . . "***

An octave can be thought of as an interval between two notes when one is half or twice the frequency of vibration of the other. A typical adult male speaking voice has a frequency of 110 Hz, an adult female 220 Hz. (Hz = Hertz, and is a unit of measurement equal to 1 cycle of vibration per second.)

p. 19 **". . . we don't think babies are cute because they are intrinsically or objectively cute . . ."**

Cuteness is a product of the perception and interpretation of a mind, human or otherwise. It is not an inherent property of any object.

p. 20 **"What distinguishes us most is one thing no other animals do: *art*."**

I leave aside the controversial question of elephant art and other similar demonstrations. When given brushes, paints, and some instruction, elephants will paint on canvases, and generally their paintings are best described as "abstract." A few paintings resemble flowers, but it is difficult to determine the extent to which these flowers were more the product of zealous human instructors. Given appropriate instruments, elephants will also make sound together in something that resembles music, and which my colleague Ani Patel has studied, noting that they can maintain a remarkably steady beat. Without the interference of humans, there is no evidence that elephants or other animals would engage in these activities on their own, and calling them artistic expression strikes me as a case of gross anthropomorphism combined with wishful thinking.

p. 23 **"Traditionally poetry has been discussed in terms of these forms (rhyming patterns, metrical patterns, number of lines)."**

In 2008, Helen Vendler adopted a more flexible attitude toward form—form with her is virtually synonymous with style:

> Each poem is a new personal venture made functional by technical expertise; the poet's moral urgency in writing is as real, needless to say, as his technical skill, but moral urgency alone never made a poem. On the other hand, technical expertise alone does not suffice, either. Form is the necessary and skilled embodiment of the poet's moral urgency, the poet's method of self-revelation.

Vendler, H. (2008, January–February). Poems are not position papers. *Harvard Magazine* 25.

p. 24 **"In defending poetry not of the ivory tower sort, he [John Barr] writes: . . ."**

Barr, J. (2007). Is it poetry or is it verse? Poetry Foundation. Retrieved December 1, 2007, from http://www.poetryfoundation.org/journal/feature.html ?id=178645, chap. 1.

p. 26 **" 'Poems are hypothetical sites of speculation, not position papers.' "**

Vendler, H. (2008, January–February). Poems are not position papers. *Harvard Magazine* 25.

p. 34 **". . . a Lakoffian metaphor . . ."**

Lakoff, G. (1987). *Women, Fire and Dangerous Things: What Categories Reveal About the Mind.* Chicago: University of Chicago Press.

p. 38 **" 'Art, in my opinion, has remained a key to survival.' "**

Read, H. (1955). *Icon and Idea.* Cambridge, MA: Harvard University Press.

"Drawings, paintings, sculpture, poems, and song allow the creator to represent an object in its absence . . ."

Here I am borrowing liberally from: Storr, A. (1992). *Music and the Mind.* New York: Ballantine Books, p. 2.

CHAPTER 2

p. 45 **"The surprise, predawn attack was a gruesome innovation in prehistoric warfare."**

I don't mean to imply that all invasions were motivated by unbridled aggression—they were often the result of the same sorts of forces that cause conflict today, such as unequal distribution of resources. Two tribes may have coexisted peacefully for centuries, when one loses its source of water—a stream could dry up. They will die without water, and the neighboring tribe is unwilling to share theirs. The waterless tribe has to choose between dying and attacking the selfish neighbors.

p. 50 **"Rhythm in music provides the input to the human perceptual system that allows for the prediction and synchronization of different individuals' behaviors."**

Condon, W. S. (1982). Cultural microrhythms. In *Interaction Rhythms,* edited by M. Davis. New York: Human Sciences Press, pp. 53–77.

pp. 50–51 **"Singing together releases oxytocin, a neurochemical now known to be involved in establishing bonds of trust between people."**

Kosfeld, M., M. Heinrichs, P. Zak, U. Fischbacher, and E. Fehr (2005). Oxytocin increases trust in humans. *Nature* 435: 673–676.

pp. 51–52 **" 'Without rhythmical coordination of the muscular effort . . . famous monuments could not have been built.' "**

McNeill, W. (1995). *Keeping Together in Time: Dance and Drill in Human History.* Cambridge, MA: Harvard University Press, p. 55.

p. 53 **"Track lining songs . . ."**

I thank Dennis Drayna for this example and its wording.

p. 54 **". . . psychologist Mihaly Csikszentmihalyi."**

Pronounced Mee-high Cheek-sent-mee-high-yee.

p. 55 **"What I remember now, years afterwards, is that I rather liked strutting around . . ."**

McNeill, W. (1995). *Keeping Together in Time: Dance and Drill in Human History.* Cambridge, MA: Harvard University Press, p. 2.

". . . some people . . . enjoy eating dirt."

This practice is known as *geophagy,* but the benefits described here are made up.

p. 56 **"The cats with this mutation were less likely to get sick or to spread**

disease to their offspring, facilitating this mutation's rapid spread through the genome."

This example and much of its wording comes from my colleague Jim Plamandon, to whom I am grateful.

". . . those who enjoyed singing, dancing, and marching together so much that they were drawn to it, attracted to it, and practiced it for thousands of hours were those who were the victors in any battles in which such drill conferred an advantage."

Of course in many cases, people were conscripted to service and forced to march. But the example still works; those who derived no enjoyment from such drill were not likely to practice on their own time, and so didn't become as expert. Further, those who enjoyed it were more likely to be good at it, and to demonstrate skill and enthusiasm on the battlefield. In fact, it has been noted that natural selection could conceivably, in the long run, tend to favor aggressive murderous psychopaths to the extent that they are able to wipe out passive, peace-loving peoples.

p. 57 **" 'Away from the cover of trees, safety can only be found in numbers. . . ."**

Mithen, S. (2005). *The Singing Neanderthals: The Origins of Music, Language, Mind and Body*. Cambridge, MA: Harvard University Press, p. 126.

p. 59 **"Those individuals who were better able to predict the behavior of others because they could 'read their minds' would have had a competitive advantage within the group."**

A point made by Mithen using similar wording.

Mithen, S. (2005). *The Singing Neanderthals: The Origins of Music, Language, Mind and Body*. Cambridge, MA: Harvard University Press, p. 128.

p. 62 **"The rappers . . . 'interpret and articulate the fears, pleasures, and promises of young black women whose voices have been relegated to the margins of public discourse.' "**

Rose, Tricia. (1994). *Black Noise: Rap Music and Black Culture in Contemporary America*. Hanover, NH: Wesleyan University Press, p. 146.

p. 63 **"One band, the Plastic People of the Universe (PPU), . . . is widely regarded as having spurred a revolution in Czechoslovakia."**

This story and the quotes from Ivan Bierhanzl come from the review of Tom Stoppard's play *Rock 'n Roll* appearing in *The New York Times*. Parales, J. (November 11, 2007). Rock 'n Revolution. *The New York Times*.

p. 69 **". . . the protests, women's lib, and improved race relations were all bound up into one big cause, into us against them."**
And it seemed so simple: if you had long hair, you were for these things. If you had short hair, many long-haired people assumed you were in favor of napalming innocent babies in a country that we weren't even at war with (Cambodia), you believed the white race to be superior to others, and you hated rock music.

p. 72 **"'As I said earlier, good music can leap over language boundaries, over barriers of religion and politics and hit someone's heartstrings somehow.'"**
See Chapter 1.
"Bruce Cockburn wrote an antiwar song, 'If I Had a Rocket Launcher.'"
The quotes from Cockburn come from an article appearing in the *Washington Post*. Harrington, R. (October 19, 1984). The Long March of Bruce Cockburn: From Folkie to Rocker, Singing About Injustice. *Washington Post*.

CHAPTER 3

p. 87 **". . . Log Blues."**
Another famous Ren and Stimpy song is the "Happy Happy Joy Joy Song." It is just as ebullient, but I reprinted the Log song because it's sillier and connects to the Slinky song a few lines down.

pp. 88–89 **". . . the ancient Greeks . . . used harp music to ease the outbursts of people with mental illnesses."**
Shapiro, A. (1969). A pilot program in music therapy with residents of a home for the aged. *The Gerontologist* 9(2): 128–133.

p. 89 **". . . the brain has been shaped by evolution and adaptations that arose independently of one another to solve specific problems."**
Marcus, G. (2008). *Kluge: The Haphazard Construction of the Human Mind*. New York: Houghton-Mifflin.
". . . adaptations such as the ability to anticipate the future, solve

puzzles, distinguish animate from inanimate objects, identify friends and enemies, and avoid being manipulated or deceived."

Huron, D. (2005). The plural pleasures of music. In *Proceedings of the 2004 Music and Music Science Conference,* edited by J. Sundberg and W. Brunson. Stockholm: Kungliga Musikhögskolan & KTH, pp. 1–13.

p. 90 **"Recall Daniel Dennett's argument that we don't find babies cute because they are *intrinsically* cute . . ."**

See Chapter 1.

"When we find something pleasurable or displeasurable, it is often because tens of thousands of years of brain evolution have *selected* for those emotions. . . ."

Jamshed Bharucha, a music cognition professor at Tufts University and former editor of the journal *Music Perception,* adds: "Many experiences of pleasure and displeasure, including disgust, outrage, liking, pleasantness, are the result of cultural familiarity or violation thereof. In some cultures, grasshoppers and dogs are considered delicious; in other cultures, the idea of eating them would be considered revolting. People familiar with the operatic voice love it. Others hate it. I have come across teachers of Western classical voice who find the Karnatic classical voice to be nasal and ugly—it goes against everything they teach. Many highly trained classical musicians have a hard time appreciating the classical musics of other cultures. I have found some of the most skilled Indian classical musicians (of older generations, who have not had early exposure to Western music) just don't get the big deal about Beethoven. Ditto in reverse. I always have been amazed at how indifferent so many skilled musicians are to other forms of music. This is not true of all musicians, but certainly of many."

p. 91 **" . . .'Suspicious Minds'. . ."**

James, M. (1956). Suspicious minds [Recorded by Elvis Presley]. On *Suspicious Minds* [45rpm record]. RCA. (1969).

"Suspicious Minds" has also been recorded by Fine Young Cannibals, Dwight Yoakam, Robbie Williams, the punk band Avail, and many others.

p. 91–92 **"Nature doesn't build mental devices whose purpose isn't related to adaptive fitness."**

Huron, D. (2005). The plural pleasures of music. *Proceedings of the 2004*

Music and Music Science Conference, edited by J. Sundberg and W. Brunson. Stockholm: Kungliga Musikhögskolan & KTH, p. 2.

p. 92 **"Although there do exist discrete 'pleasure centers' in the brain, dozens of neurotransmitters and brain regions contribute to feelings of pleasure."**

These two sentences are nearly direct quotes from: Huron, D. (2005). The plural pleasures of music. *Proceedings of the 2004 Music and Music Science Conference,* edited by J. Sundberg and W. Brunson. Stockholm: Kungliga Musikhögskolan & KTH, p. 2.

p. 94 **"In one published study on music therapy, a group of Korean researchers took stroke survivors and gave them an eight-week program of physical therapy that involved synchronized movements to music."**

Jeong, S. and M. T. Kim. (2007). Effects of a theory-driven music and movement program for stroke survivors in a community setting. *Applied Nursing Research* 20(3): 125–31.

p. 95 **"In fact, in the long run, she will tend to get 25 percent of them [the cards] right."**

This is because for any given "trial" (that is, each time your friend tries to guess the suit of a card) there are four possibilities: hearts, clubs, diamonds, and spades. On Trial 1 you might be looking at a spade and your friend might guess heart. On Trial 2 you might be looking at a heart and your friend might guess diamond. But on the average, your friend's guesses will align with yours one out of four times; if she is truly guessing at random, that is, if she has no idea. For that matter, if you friend tries to be stubborn and just says "hearts" on *every single trial,* even though they can't *all* be hearts, she will still be right on 25 percent of the trials.

p. 98 **". . . oxytocin . . ."**

$C_{43}H_{66}N_{12}O_{12}S_2$. It is produced in the hypothalamus.

"Serum concentrations of oxytocin increased significantly [in people who had been given singing lessons]."

Grape, C., M. Sandgren, L. O. Hansson, M. Ericson, and T. Theorell. (2003). Does singing promote well-being? *Integrative Physiological & Behavioral Science* 38(1): 65–74.

". . . oxytocin has just been found to increase trust between people."

Kosfeld M., M. Heinrichs, P. Zak, U. Fischbacher, and E. Fehr. (2005). Oxytocin increases trust in humans. *Nature* 435: 673–676.

"Why oxytocin is released when people sing together is probably related evolutionarily to the social bonding function of music . . ."

Freeman, W. J. (1995). *Societies of Brains: A Study in the Neuroscience of Love and Hate.* Hillsdale, NJ: Erlbaum.

pp. 98–99 **"Several recent studies show that IgA levels increased following various forms of music therapy."**

Charnetski, C. J., G. C. Strand, M. L. Olexa, L. J. Turoczi, and J. M. Rinehart. (1989). The effect of music modality on immunoglobulin A (IgA). *Journal of the Pennsylvania Academy of Science* 63: 73–76.

Kuhn, D. (2002). The effects of active and passive participation in musical activity on the immune system as measured by salivary immunoglobulin A (SIgA). *Journal of Music Therapy* 39(1): 30–39.

McCraty, R., M. Atkinson, G. Rein, and A. D. Watkins. (1996). Music enhances the effect of positive emotional states on salivary IgA. *Stress Medicine* 12(3): 167–175.

McKinney, C. H., M. H. Antoni, M. Kumar, F. C. Tims, and P. McCabe. (1997). Effects of guided imagery and music (GIM) therapy on mood and cortisol in healthy adults. *Health Psychology* 16(4): 390–400.

McKinney, C. H., F. C. Tims, A. M. Kumar, M. Kumar. (1997). The effect of selected classical music and spontaneous imagery on plasma beta-endorphin. *Journal of Behavioral Medicine* 20(1): 85–99.

Rider, M. S., and J. Achterberg. (1989). Effect of music-assisted imagery on neutrophils and lymphocytes. *Applied Psychophysiology and Biofeedback* 14(3): 247–257.

Tsao, J., T. F. Gordon, C. Dileo, and C. Lerman. (1999). The effects of music and biological imagery on immune response. *Frontier Perspectives* 8: 26–37.

p. 99 **"In another study, levels of melatonin, norepinephrine, and epinephrine increased during a four-week course of music therapy . . ."**

Kumar, A. M., F. Tims, D. G. Cruess, M. J. Mintzer, G. Ironson, D. Loewenstein, et al. (1999). Music therapy increases serum melatonin levels in patients with Alzheimer's disease. *Alternative Therapies in Health and Medicine* 5(6): 49–57.

"Melatonin . . ."

$C_{13}H_{16}N_2O_2$.

". . . some researchers believe that it [melatonin] increases cytokine production, which in turn signals T-cells to travel to the site of an infection."

Carrillo-Vico, A., R. J. Reiter, P. J. Lardone, J. L. Herrera. R. Fernández-Montesinos, J. M. Guerrero, et al. (2006). The modulatory role of melatonin on immune responsiveness. *Current Opinion in Investigating Drugs* 7(5): 423–431.

"Serotonin levels were shown to increase in real time during listening to pleasant, but not unpleasant music."

Evers, S., and B. Suhr. (2000). Changes of the neurotransmitter serotonin but not of hormones during short time music perception. *European Archives of Psychiatry and Clinical Neuroscience* 250(3): 144–147.

"Techno music increased levels of plasma norepinephrine (NE), growth hormone (GH) . . ."

Gerra, G., A. Zaimovic, D. Franchini, M. Palladino, G. Giucastro, N. Reali, et al. (1998). Neuroendocrine responses of healthy volunteers to "techno-music": Relationships with personality traits and emotional state. *International Journal of Psychophysiology* 28(1): 99–111.

". . . rock music was shown to cause decreases in prolactin . . . a hormone associated with feeling good."

Möckel, M., L. Röcker, T. Stork, J. Vollert, O. Danne, H. Eichstädt, et al. (1994). Immediate physiological responses of healthy volunteers to different types of music: Cardiovascular, hormonal and mental changes. *European Journal of Applied Physiology* 68(6): 451–459.

p. 103 **". . . *Sweet Anticipation* . . ."**

Huron, D. (2006). *Sweet Anticipation: Music and the Psychology of Expectation.* Cambridge, MA: MIT Press.

See also the excellent book review:

Stevens, C., and T. Byron. (2007). Sweet anticipation: Music and the psychology of expectation. [Review of the book *Sweet Anticipation: Music and the Psychology of Expectation*]. *Music Perception* 24(5): 511–514.

p. 105 **"In a paper published a few years ago in the journal *Music Perception* . . ."**

Vines, B. W., R. L. Nuzzo, and D. J. Levitin. (2005). Analyzing temporal dynamics in music: Differential calculus, physics, and functional data analysis techniques. *Music Perception* 23(2): 137–152.

p.106 **". . . tension tends to build up during music to a peak, after which the tension is released and subsides, often rapidly."**

It is true that composers sometimes flout these conventions and write pieces with no tension, pieces that end on tension rather than resolving, and so on. But these are relatively uncommon compared to what typically occurs; indeed, their relative rarity is what gives them their power to surprise.

"In performances of Indian classical music . . ."

The wording for this section comes from Jamshed Bharucha.

p. 107 **". . . 'Over the Rainbow' . . ."**

Arlen, H., and E. Y. Harburg. (1939). Over the rainbow [Recorded by Judy Garland]. On *Over the Rainbow* [LP]. Pickwick Records.

". . . 'She Loves You' . . ."

Lennon, J., and P. McCartney. (1963). She loves you [Recorded by The Beatles]. On *She Loves You* [45rpm record]. London: Parlophone Records.

p. 109 **"'I define joy . . . as a sustained sense of well-being and internal peace . . .'"**

Oprah Winfrey. (n.d.). Retrieved March 7, 2008, from http://en.wikiquote.org/wiki/Oprah_Winfrey, accessed March 7, 2008.

CHAPTER 4

p. 125 **"In many of the places I've worked, music has been there as a soundtrack to help the employees get through their day."**

In the 1970s (some years before my chef's job at Sambo's), I worked as a dishwasher at Scoma's, a seafood restaurant in Sausalito, California. There, we often listened to the song "Hard Work" by John Handy. The manager would play it every night at the beginning of the shift as we all scurried to make the restaurant ready for the frenzied, every-night-a-capacity-crowd dinner rush. The stress of getting all the food prepared, pots and pans clean and oiled, tables set, menus printed, and so on, was greatly relieved by music that the manager piped throughout the back room and entire restaurant over the PA.

When this song came on, people's tense shoulders would drop an inch, their footsteps become lighter, their actions become more fluid and graceful. The heavy beat of the song and its I-bVII vamp give it a sense of gravity, but the performance is so *jovial* and ebullient that it feels almost heliumlike in its ability to elevate drudgery and tension to purposefulness and a confidence that everything will work out right. So many songs with vamps set up a groove that in turn conveys a sense of timelessness—we forget about the clock and feel that if something goes wrong, no problem—we can just do it over again. Any thoughts that we may run out of time are vanquished by the alternate universe of the song, where the beat is marked at regular, rhythmic intervals and the song moves unflinchingly forward, but ordinary "world time" seems to stand still.

Handy, J. (1976). Hard work. On *Hard Work* [LP]. Impulse! Records

p. 126 **"Mothers from every culture sing to their infants."**

For more on the evolution of behavior, see the excellent books by Sarah Blaffer Hrdy: Hrdy, S. B. (1981). *The Woman That Never Evolved.* Cambridge: Harvard University Press.

———. (1981). *Mother Nature: A History of Mothers, Infants, and Natural Selection.* New York: Pantheon.

p. 131 **". . . 'God Bless America.' "**

Irving Berlin, a Jewish immigrant from Siberia, wrote the song in 1918. In 1938 he revised it, and it was reintroduced on Armistice Day that year, sung by Kate Smith. There have been movements over the years to adopt it officially as America's national anthem. Woody Guthrie reportedly wrote "This Land Is Your Land" as a musical reply to "God Bless America." ASCAP, the composers' rights agency, reports that "God Bless America" was by far the most played song in the months following 9/11.

p. 133 **"Sorrow does have an evolutionary purpose . . ."**

Brean, J. (December 8, 2007). Chemicals play key role in a person's appreciation of sad music, expert says. [Electronic version]. *National Post.* Retrieved March 5, 2008, from http://www.nationalpost.com/Story.html?id=154661.

CHAPTER 5

p. 140 **"The reason that most frogs synchronize their calls is that it makes it more difficult for predators to locate them . . ."**

Tuttle, M. D., and M. J. Ryan, (1982). The role of synchronized calling, ambient light, and ambient noise, in anti-bat-predator behavior of a treefrog. *Behavioral Ecology and Sociobiology* 11: 125–131.

p. 141 **"survival is only enhanced by sorting out fact from fiction . . ."**

Of course, organisms can "get ahead" by telling lies, Huron told me. So, *deceptive* communication can enhance survival of the deceiver. The point is not that communication should be truthful, but that organisms will have a selective advantage if they can decipher fact from fiction. One can easily imagine an arms race of deceivers trying to stay one step ahead of a developing ability to spot deceit, or vice-versa, and this does occur in the animal kingdom.

"Music's direct and preferential influence on emotional centers of the brain and on neurochemical levels supports this view [that music and brains co-evolved]."

See Chapter 4.

p. 143 **"By seven months, infants can remember music for as long as two weeks . . ."**

Saffran, J. R., M. M. Loman, and R. R. Robertson. (2000). Infant memory for musical experiences. *Cognition* 77(1): B15–B23.

". . . mother-infant vocal interactions exhibit striking similarities across a wide range of cultures."

Trehub, S. (2003). The developmental origins of musicality. *Nature Neuroscience* 6(7): 669–673.

And summarized in: Cross, I. (in press). The evolutionary nature of musical meaning. *Musicae Scientiae.*

See also, for related and relevant ideas: Cross, I. (2007). Music and cognitive evolution. In *Handbook of Evolutionary Psychology,* edited by R. I. Dunbar and L. Barrett. Oxford, UK: Oxford University Press, pp. 649–667.

Cross, I. (in press). Music as a communicative medium. In *The Prehistory of Language* (Vol. 1), edited by C. Knight and C. Henshilwood. Oxford, UK: Oxford University Press.

Cross, I. (in press). Musicality and the human capacity for culture. *Musicae Scientiae.*

"Mothers also use these musiclike vocalizations to direct their infants' attention . . ."

Dissanayake, E. (2000). Antecedents of the temporal arts in early mother-infant interactions. In *The Origins of Music,* edited by N. Wallin, B. Merker, and S. Brown. Cambridge, MA: MIT Press, pp. 389–407.

Gratier, M. (1999). Expressions of belonging: The effect of acculturation on the rhythm and harmony of mother-infant interaction. Musicae Scientiae Special Issue: 93–112.

p. 146 **". . . some mice might have stumbled upon the fact that if *they* made low-pitched sounds with their throats and mouths, it might serve to intimidate other mice . . ."**

Owings and Morton call this "expressive size symbolism." Owings, D. H., and E. S. Morton. (1998). *Animal Vocal Communication: A New Approach.* Cambridge, UK: Cambridge University Press.

See also: Cross, I. (in press). The evolutionary nature of musical meaning. *Musicae Scientiae.*

p. 147 **"'Vicarious musical pleasure . . . seems to put a damper on musical self-expression.'"**

Robison, P. (n.d) *Blackwalnut Interiors.* Unpublished manuscript. I am grateful to Paula Robison's grandson Toby Robison for providing this.

p. 152 **"In the 1930s, Albert Lord and Milman Parry recorded folk songs in the mountains of (then) Yugoslavia . . ."**

Lord, A. B. (1960). *The Singer of Tales.* Cambridge, MA: Harvard University Press.

p. 152–53 **"Some of them memorize their songs with very high accuracy . . ."**

Lord, A. B. (1960). *The Singer of Tales.* Cambridge, MA: Harvard University Press.

p. 153 **"The Gola of West Africa place a particularly high value on the preservation and transmission of tribal history."**

D'Azevedo, W. L. (1962). Uses of the past in Gola discourse. *Journal of African History* 3: 11–34.

p. 154 **"Oliver's mind had brought up Mahler's song of mourning for the death of children . . ."**

Sacks, O. (2007). *Musicophilia: Tales of Music and the Brain*. New York: Knopf, p. 280.

"'All this time, I still remember everything you said'"

Banks, T., P. Collins, and M. Rutherford. (1986). In too deep [Recorded by Genesis]. On *Invisible Touch* [CD]. Virgin Records.

"'I remember the smell of your skin . . .'"

Adams, B., and R. Lange. (1993). Please forgive me [Recorded by Bryan Adams]. On *So Far So Good* [CD]. A&M Records.

p. 156 **". . . mutually reinforcing, multiple constraints of songs are crucially what keeps oral traditions stable over time."**

Wallace, W. T., and D. C. Rubin. (1988). "The wreck of the old 97": A real event remembered in song. In *Remembering Reconsidered: Ecological and Traditional Approaches to the Study of Memory*, edited by U. Neisser and E. Winograd. Cambridge, UK: Cambridge University Press, pp. 283–310.

p. 159 **". . . literal recall is seldom important."**

Bartlett, F. C. (1932). *Remembering: A Study in Experimental and Social Psychology*. London: Cambridge University Press.

p. 160 **"It may also seem . . . that your brain is *not* generating all the possible rhymes for a forgotten word, but research has shown that this is in fact what's happening . . ."**

Kintsch, W. (1988). The role of knowledge in discourse comprehension: A construction-integration model. *Psychological Review* 95(2): 163–182.

Schwanenflugel, P. J., and K. L. LaCount. (1988). Semantic relatedness and the scope of facilitation for upcoming words in sentences. *Journal of Experimental Psychology: Learning, Memory, and Cognition*, 14: 344–354.

p. 163 **". . . when singers of a given tradition are asked to write a *new* ballad . . . they tend to employ all of the same tools . . ."**

Wallace, W. T, and D. C. Rubin. (1991). Characteristics and constraints in ballads and their effects on memory. *Discourse Processes* 14: 181–202.

". . . they changed twenty-four words in 'The Wreck' to eliminate assonance, alliteration, and rhyme."

Wallace, W. T., and D. C. Rubin. (1988). "The wreck of the old 97": A real event remembered in song. In *Remembering Reconsidered: Ecologi-*

cal and Traditional Approaches to the Study of Memory, edited by U. Neisser and E. Winograd. Cambridge, UK: Cambridge University Press, pp. 283–310.

p. 165 **". . . there are strong cultural pressures to recall such material accurately or not at all."**

Rubin, D. C. (1995). *Memory in Oral Traditions: The Cognitive Psychology of Epic, Ballads, and Counting-out Rhymes.* New York: Oxford University Press, p 179.

p. 166 **"Insight into the matter [theory of multiple, reinforcing constraints] comes from another very clever experiment by Rubin."**

Rubin D. C. (1977). Very long-term memory for prose and verse. *Journal of Verbal Learning and Verbal Behavior* 16(5): 611–621.

"Rubin asked fifty people to recall the words of the Preamble to the United States Constitution . . . "

"We the People of the United States, in Order to form a more perfect Union, establish Justice, insure domestic Tranquility, provide for the common defence, promote the general Welfare, and secure the Blessings of Liberty to ourselves and our Posterity, do ordain and establish this Constitution for the United States of America."

"This of course does not have music . . . "

My graduate student, Mike Rud (considered one of the best jazz guitarists in Canada) tells me: "As a very young Canadian, my first exposure to the Preamble of the U.S. Constitution was the fantastic *Schoolhouse Rock* version from ABC's Saturday morning cartoons. A recent resurgence of interest in *Schoolhouse Rock* lead to its reissue on DVD. I find for memorizing this passage, this funky melody is of more help than the beautiful but rather convoluted syntax of the original. The listener is kept waiting for well over thirty words after the sentence's subject before hearing the verb! But the melody's unique constraints really help a kid remember some otherwise difficult-to-parse prose."

p. 167 **". . . these rhythmic units usually coincide with the units of meaning in oral traditions."**

Rubin, D. C. (1995). *Memory in Oral Traditions: The Cognitive Psychology of Epic, Ballads, and Counting-out Rhymes.* New York: Oxford University Press, p. 179.

Rubin also cites: Bakker, E. J. (1990). Homeric discourse and enjambement: A cognitive approach. *Transactions of the American Philolological Association* 120: 1–21.

Lord, A. B. (1960). *The Singer of Tales*. Cambridge, MA: Harvard University Press.

Parry, M. (1971a). Homeric formulae and Homeric metre. In *The making of Homeric Verse: The Collected Papers of Milman Parry,* edited and translated by A. Parry. Oxford, UK: Oxford University Press, pp. 191–239. (Original work published 1928.)

Parry, M. (1971b). The traditional epithet in Homer. In *The making of Homeric verse: The Collected papers of Milman Parry,* edited and translated by A. Parry. Oxford, UK: Oxford University Press, pp. 1–190. (Original work published 1928.)

p. 170 **"The reality of these chunks has been demonstrated many times . . ."**

Another good example of using poetics to remember was provided to me by Jamshed Bharucha. An Indian man, Rajan Mahadevan, was at one point listed in the *Guinness Book of World Records* for being able to recite thirty thousand plus digits of Pi from memory. He showed Bharucha how he uses chunking, meter, and rhythm. His digit span and spatial memory are not much different from average, but through chunking, poetics, and lots and lots of practice, he got to this huge number.

p. 171 **". . . it takes college undergraduates much longer to say what letter comes just before *h, l, q,* or *w* than before *g, k, p,* and *v.*"**

Klahr, D., W. G. Chase, and E. A. Lovelace. (1983). Structure and process in alphabetic retrieval. *Journal of Experimental Psychology: Learning, Memory, and Cognition* 9(3): 462–477.

". . . some professional musicians and Shakespearean actors do indeed have perfect recall for a memorized string and can begin anywhere . . ."

Oliver, W. L., and K. A. Ericsson. (1986). Repertory actors' memory for their parts. In *Proceedings of the Eighth Annual Conference of the Cognitive Science Society*. Hillsdale, NJ: Erlbaum, pp. 399–406.

p. 172 **" 'I will sing it and you tell me when the demon you want has his name mentioned.' "**

This is reported as a personal communication (occurring in November 1991) from Kapferer to David Rubin.

Rubin, D. C. (1995). *Memory in Oral Traditions.* New York: Oxford University Press, p. 190.

p. 173 **". . . words containing a long-short-long or short-short-short syllabic structure can't be used in Homeric epic . . ."**

Rubin, D. C. (1995). *Memory in Oral Traditions.* New York: Oxford University Press, p. 198.

". . . in the Zoroastrian tradition . . ."

I thank Jamshed Bharucha for contributing the passage about Zoroastrian prayers.

p. 174 **". . . 'music and singing carrying obscene content . . .' "**

On the Correspondence of Music, Musical Instruments and Singing to the Norms of Islam. (2005). Retrieved March 6, 2008, from http://umma.ws/Fatwa/music/.

p. 175 **". . . the Talmud *is*—a record of what were essentially judicial proceedings and deliberations about what precisely the oral teachings were . . ."**

I've somewhat simplified the story here, because I'm not so concerned in this book with details of the Torah and its transmission, but rather, with the fact that it was set to music and forms a nice example of a knowledge song. But I'll expand a bit here. According to traditional rabbinic sources, the entire Torah was given to Moses by God, and it was given in two parts: the Torah proper (which was allowed to be written down) and a system of commentaries and emendations (known as the "Oral Torah"). The Oral Torah—according to the rabbis—was the part that was not written down for one thousand years, and about which there were many debates. It is possible, however, that the so-called "Written Torah"—what we call the Five Books of Moses (the first five chapters of the Old Testament: Genesis, Exodus, Numbers, Chronicles, and Deuteronomy)—was not actually written down *either* during hundreds of years; we have only the rabbis' teachings on this. The earliest known/preserved written documents, the Dead Sea Scrolls, have been carbon dated to be no older than the second century B.C.E., so we don't have any independent confirmation that the Written Torah was written down any earlier than the Oral Torah. There are debates on both sides, and at this point, the debates have not been resolved by any physical evidence.

p. 176 **". . . if melodies can change, so can words."**

One way we infer this is by the discovery of Jews in Ethiopia in the 1980s who had been cut off from contact with other Jews for two thousand years or so, and who are believed to be descended from a liaison between the queen of Sheba and King Solomon. DNA studies have failed to show a genetic descent, and the dominant scholarly view is that contemporary Ethopian Jews are descended from local converts. Regardless, when discovered, they believed that they were the only Jews in the world. They had Torah scrolls and observances similar to those of contemporary Jews, but they did not celebrate the post-biblical holidays of Purim or Hanukkah, having been cut off from the rest of world Jewry *after* the establishment of those holidays. Many of the melodies they sang for religious songs, Psalms, and Torah were different from any that are currently sung today. Some believe that these melodies may be closer to the original melodies sung by King Solomon himself and, hence, closer to the melodies sung by King David, Moses, and biblical-era Jews. The point is that the drift in melodies between these two groups, whose separation over two millenia seems difficult to dispute, is evidence that the melodies can and do change over time, and if melodies can, so can words.

p. 178 **". . . [the writer] weaves into the message two well-known Old Testament references."**

Performed by Hank Williams, Aubrey Gass, Tex Ritter, and others.

Gass, A. (1949). Dear John [Recorded by Hank Williams]. On *Dear John* [45rpm record]. MGM Records. (1951).

p. 180 **". . . there are manifest cognitive benefits that are conferred to the group-as-a-whole . . . when people sing together."**

In addition to those references previously cited, for a computer science/artificial intelligence perspective, see also:

Gill, S. P. (2007). Entrainment and musicality in the human system interface. *AI & Society* 21(4): 567–605.

p. 181 **"No single ant 'knows' that the hill needs to relocate . . . but the actions of tens of thousands of ants result in the hill being moved . . ."**

See for example: Gordon, D. M. (1999). *Ants at Work: How an Insect Society Is Organized*. New York: The Free Press.

Johnson, S. (2001). *Emergence: The Connected Lives of Ants, Brains, Cities, and Software.* New York: Scribner.
Strogatz, S. H. (1994). *Nonlinear Dynamics and Chaos: With Applications to Physics, Biology, Chemistry and Engineering.* Cambridge, MA: Perseus Books.
Strogatz, S. H. (2003). *Sync: How Order Emerges from Chaos in the Universe, Nature, and Daily Life.* New York: Hyperion.
Wiggins, S. (2003). *Introduction to Applied Nonlinear Dynamical Systems and Chaos.* New York: Springer-Verlag.

". . . *nonlinear dynamical systems* . . . even the faddish propagation of hit songs . . ."
The art of *mixing* a popular song has nonlinear components, according to producer and pundit Sandy Pearlman, as instrumental parts interact with one another and with signal processing devices in ways that are too difficult to easily predict or characterize.

The mathematics involved for calculating a single interaction is usually nothing more complicated than what Newton had available in his time, but he didn't have the computational capacity we have now to model all the possible interactions. (Without a computer, Newton couldn't cope with all the calculations necessary to characterize three planets, let alone fifty thousand ants—there are just too many computations and the calculation grows extremely fast with the number of constituents.)

p. 184 **"This trade-off is itself nonlinear and dynamic, changing throughout the course of a performance . . ."**
I thank my colleague Frederic Guichard for this formulation.

"'Words are the children of reason and, therefore, can't explain it [music].'"
Bill Evans Quotes. (n.d.). Retrieved March 7, 2008, from http://thinkexist.com/quotes/bill_evans/.

CHAPTER 6

p. 191 **". . . the archaeological record suggests that their [Neanderthal] burials were an accidentally adopted behavior for hygienic reasons . . ."**
In spite of earlier misleading reports of Neanderthal "bear cults,"

burials, and so on. there is in fact no substantial evidence that they had any symbolic behavior or produced any symbolic objects. The burials may simply have been a way of discouraging hyenas from invading their camps. The actual data are slightly more ambiguous than I suggest here, the interested reader might refer to: Mithen, S. (2001). The evolution of imagination: An archaeological perspective. *SubStance* 30(1&2): 28–54.

p. 192 **". . . no known human culture lacks religion."**

In contemporary society we see that not everyone subscribes to religious beliefs, but this is a relatively new state, as a consequence of greater freedoms of thought in democratic societies. In the old days, if you didn't believe in the state- or community-sanctioned religion, you were typically killed.

". . . religion is more than a meme . . ."

See Dawkins, R. (1976). *The Selfish Gene.* Oxford, UK: Oxford University Press.

". . . anything that is universal to human culture is likely to contribute to human survival."

Durkheim, É. (1965). *The Elementary Forms of the Religious Life,* translated by J. W. Swain. New York: The Free Press, p. 87. (Original work published 1912.)

p. 193 **"Rituals involve repetitive movement."**

This is notwithstanding the advice of the Bigelow Tea Company, written on the outside of their tea bags (one of which I am steeping in a cup right now), to "indulge in the ancient ritual of drinking green tea." Drinking tea may be a habit, it may even be part of some ceremonies, but it doesn't qualify as a *ritual* in the strict sense.

". . . Rappaport defined ritual as 'acts of display through which one or more participants transmit information . . .' "

Rappaport, R. A. (1971). The sacred in human evolution. *Annual Review of Ecology and Systematics* 2, p. 25.

p. 196 **"Most children enter a stage of development . . . in which they show phases of ritual behaviors . . ."**

Boyer, P. and P. Liénard. (2006). Why ritualized behavior? Precaution Systems and action parsing in developmental, pathological and cultural rituals. *Behavioral and Brain Sciences* 29(6): 1–56.

p. 197 **". . . many children connect their ad hoc rituals to the supernatural or to magic . . ."**

See: Evans, D. W., M. E. Milanak, B. Medeiros, and J. L. Ross. (2002). Magical beliefs and rituals in young children. *Child Psychiatry and Human Development* 33(1): 43–58.

p. 198 **". . . ritual adds a sense of order, constancy, and familiarity . . ."**

Dulaney, S., and A. P. Fiske. (1994). Cultural rituals and obsessive-compulsive disorder: Is there a common psychological mechanism? *Ethos* 22(3): 243–283.

Zohar, A. H., and L. Felz. (2001). Ritualistic behavior in young children. *Journal of Abnormal Child Psychology* 29(2): 121–128.

"Oxytocin . . . has been found to be connected to the performance of ritual . . ."

Leckman, J. F., R. Feldman, J. E. Swain, V. Eicher, N. Thompson, and L. C. Mayes. (2004) Primary parental preoccupation: Circuits, genes, and the crucial role of the environment. *Journal of Neural Transmission* 111(7): 753–771.

"From an adaptation perspective, this order makes any intrusion by an outsider immediately and clearly visible."

Boyer, P., and P. Liénard. (2006). Why ritualized behavior? Precaution Systems and action parsing in developmental, pathological and cultural rituals. *Behavioral and Brain Sciences* 29(6): 10.

p. 199 **"The basal ganglia store chunks or summaries of motor behavior . . ."**

Canales, J. J., and A. M. Graybiel. (2000). A measure of striatal function predicts motor stereotypy. *Nature Neuroscience* 3(4): 377–383.

Graybiel, A. M. (1998). The basal ganglia and chunking of action repertoires. *Neurobiology of Learning and Memory* 70(1–2): 119–136.

Rauch, S. L., P. J. Whalen, C. R. Savage, T. Curran, A. Kendrick, H. D. Brown, et al. (1997). Striatal recruitment during an implicit sequence learning task as measured by functional magnetic resonance imaging. *Human Brain Mapping* 5(2): 124–132.

Saxena, S., A. L. Brody, K. M. Maidment, E. C. Smith, N. Zohrabi, E. Katz, et al. (2004). Cerebral glucose metabolism in obsessive-compulsive hoarding. *American Journal of Psychiatry* 161(6): 1038–1048.

Saxena, S., A. L. Brody, J. M. Schwartz, and L. R. Baxter. (1998).

Neuroimaging and frontal-subcortical circuitry in obsessive-compulsive disorder. *British Journal of Psychiatry* (Suppl. 35): 26–37.

p. 200 **"All higher animals have a 'security-motivation' system . . ."**

Szechtman, H., and E. Woody. (2004). Obsessive-compulsive disorder as a disturbance of security motivation. *Psychological Review* 111(1): 111–127.

"The display aspect of single rituals . . ."

Fiske, A. P., and N. Haslam. (1997). Is obsessive-compulsive disorder a pathology of the human disposition to perform socially meaningful rituals? Evidence of similar content. *Journal of Nervous and Mental Disease* 185(4): 211–222.

p. 201 **". . . the fear-security-motivation system was 'built' thousands or tens of thousands of years ago . . ."**

Boyer, P., and P. Liénard. (2006). Why ritualized behavior? Precaution systems and action parsing in developmental, pathological and cultural rituals. *Behavioral and Brain Sciences* 29(6): 1–56.

See also: Sapolsky, R. (1994). *Why Zebras Don't Get Ulcers.* New York: Henry Holt.

p. 205 **". . . the ancient *Devr* ritual of the Kotas . . ."**

Wolf, R. K. (2006). *The Black Cow's Footprint: Time, Space, and Music in the Lives of the Kotas of South India.* Urbana, IL: University of Illinois Press.

I thank Bianca Levy for finding and summarizing this ritual.

p. 206 **". . . fifth in Kyrie of the Catholic Mass . . ."**

Missa Jubilate Deo.(XI–XIII cent.) Kyrie from Mass XVI, 200. Audio and score available at http://www.adoremus.org/Kyrie.html.

p. 210 **"For the Mbuti people, the forest is benevolent and powerful, and their music is the language with which they communicate with the spirit of the forest . . ."**

Feld, S. (1996). Pygmy POP: A genealogy of schizophonic mimesis. *Yearbook for Traditional Music* 28: 1–35.

Turnbull, C. (1961). *The Forest People.* New York: Simon and Shuster.

Turnbull, C. (1965). *Wayward Servants: The Two Worlds of the African Pygmies.* Garden City, NY: Natural History Press.

"'Its [Pygmy music's] most striking features, apparently common to all groups . . ."

Cooke, P. (1980). Pygmy music. In *The New Grove Dictionary of Music and Musicians,* 15th edition p. 483.

p. 211 **"The pygmies famously resisted efforts by a few unwittingly condescending anthropologists to render them as 'primitives.'"**

Feld, S. (1996). Pygmy POP: A genealogy of schizophonic mimesis. *Yearbook for Traditional Music* 28: 1–35.

p. 215 **"'Even religion today has lost its ability to pull us out—now it's all warrior gods.'"**

Joni added: "The Genesis story, originally, was about the Earth Mother; all the primitives believe this, and I'm a primitive at heart too. It's the smarter myth, the original one. They're all myths, but of all of the myths, that's the one that is most intelligent for living on this planet: 'Earth Mother gives birth to Creation without a sire.' Then that evolved into 'Earth Mother gave birth to the planet *with* a sire,' which devolved into—these are all *de*-evolutionary—'Earth Mother is killed.' Eventually, it comes down to the last one, which is 'Father gives birth to the planet without a mother.' So here we are in this goddessless situation; out of balance, no more Mother Earth or Father Sky. Mother Earth got killed off and what we ended up with is a narcissistic, war-loving, woman-hating religion, and that's what Christianity, Islam, and Judaism are. They try to teach it otherwise, but they aren't. It's a fundamental hatred of the feminine principal and a domination of it."

p. 220 **"Every human society that historians and anthropologists have uncovered has had some form of religion . . ."**

Rappaport, R. A. (1971). The sacred in human evolution. *Annual Review of Ecology and Systematics* 2: 23–44.

pp. 222–23 **"Three emotions in particular are associated with religious ecstasy: dependence, surrender, and love."**

Otto, R. (1923). *The Idea of the Holy,* translated by J. W. Harvey. London: Oxford University Press. (Original work published 1917.)

p. 223 **"[Dependence, surrender, and love] are believed innately present in animals and human infants."**

Erikson, E. (1968). The development of ritualization. In *The Religious Situation,* edited by D. Cutler. Boston: Beacon, pp. 711–733.

Rappaport, R. A. (1971). The sacred in human evolution. *Annual Review of Ecology and Systematics* 2: 23–44.

". . . the details of musical syntax remain to be worked out . . ."

An ambitious effort to do so began with the landmark publication of:

Lerdahl, F., and R. Jackendoff. (1983). *A Generative Theory of Tonal Grammar.* Cambridge, MA: MIT Press.

p. 227 **"[Psalm 131's] Middle Eastern, minorish sound, with odd, exotic intervals, evokes stone buildings and walled cities."**
The melody of the original has been lost, but in a synagogue service in a small, remote village of Israel—Bet Shemesh—I heard it sung by Moroccan Jews who have lived as a close community for many centuries. The melody sounded to be as ancient as the song itself, a beautiful, harmonic minor with delicate ornamentation. This may have been quite close to what David himself had written.

p. 228 **"'. . . But because He designed us, He knows what *we* need.'"**
The scientist or atheist will ask then, "If God was truly not an egotist, why would he create in us a need for him?" This debate is beyond the scope of this book, but the interested reader may wish to read Daniel Dennett's *Breaking the Spell.* Dennett, D. C. (2006). *Breaking the Spell: Religion as a Natural Phenomenon.* New York: Viking.

CHAPTER 7

p. 229 **"'Romantic love songs are a sham that perpetuate a lie on unsuspecting young kids . . .'"**
The first part of this quote is from a telephone interview I conducted with Frank in 1980; the second part is from his biography: Zappa, F., and P. Occhiogrosso. (1999). *The Real Frank Zappa Book.* New York: Touchstone, p. 89.

p. 231 **"'I have had some experiences with love, or think I have . . .'"**
Vonnegut, K. (1976). *Slapstick: Or Lonesome No More!* New York: Delta Books, pp. 2–3.

p. 233 **". . . love can make you do things you might not otherwise do . . ."**
As Gabriel García Márquez writes in *Love in the Time of Cholera:* "Fermina Daza, his wife . . . was an irrational idolator of tropical flowers and domestic animals, and early in her marriage she had taken advantage of the novelty of love to keep many more of them in the house than good sense would allow." García Márquez, G. (1989). *Love in the Time of Cholera,* translated by E. Grossman. London: Penguin, p. 21. (Original work published 1985.)

p. 235 **"Because romantic love is what gets written about, talked about, filmed and sung about so much, we can temporarily forget that love comes in many different forms . . ."**

The ancient Greeks had already classified these different forms of love, and in fact, distinguished ten forms of love, and the psychologist John Alan Lee reduced these to six; see: Lee, J. A. (1976). *The Colours of Love*. Englewood Cliffs, NJ: Prentice-Hall.

Either formulation confounds the way people act (playful, generous) with the way they feel (jealous, passionate) and the underlying, unifying principles (attachment, longing, lust). In her book *Why We Love,* Helen Fisher claims that these can be reduced to three: romantic love, attachment, and lust: Fisher, H. (2004). *Why We Love: The Nature and Chemistry of Romantic Love*. New York: Henry Holt.

I think including lust as its own form of love is odd, as opposed to including it as an element in certain forms of attachment love. I find Robert Sternberg's triangular theory of love more compelling, that the various forms of love are all combinations of three basic elements: passion, intimacy, and commitment: Sternberg, R. J. (1986). A triangular theory of love. *Psychological Review* 93(2): 119–135.
Sternberg, R. J. (1988). *The Triangle of Love: Intimacy, Passion, Commitment*. New York: Basic Books.

But Sternberg's system doesn't easily account for love of an ideal (like justice), or love of country—both of which he would probably describe as "commitment and passion," but that, to me, fails to capture the different phenomenology, the different *feeling,* we have of love for our hometown than we do of love for our romantic partner. More importantly, I think they all miss the fundamental common point that love *is* caring.

p. 236 **"'Jeremiah de Saint-Amour, already lost in the mists of death . . .'"**
García Márquez, G. (1989). *Love in the Time of Cholera,* translated by E. Grossman. London: Penguin, p. 14 (Original work published 1985.)

p. 237 **". . . foodstuffs are redistributed equitably among neighboring tribes, eliminating what could be deadly food-jealousy wars."**
Ford, R. I. (1971). An ecological perspective of the eastern pueblos. In *New Perspectives on the Eastern Pueblos,* edited by A. Ortiz Albuquerque, NM: University of New Mexico Press.

p. 238 **". . . rather than building into genes and brains information that is ubiquitous and readily available in the environment, brains are configured such that they can incorporate regularities . . ."**

Deacon, T. W. (1997). What makes the human brain different? *Annual Review of Anthropology* 26: 337–357.

p. 240 **". . . males don't pair-bond with the mothers of their offspring and they don't provide paternal care."**

Diamond, J. (1997). *Why Is Sex Fun?* New York: Basic Books. Exceptions include male zebras and gorillas (who are polygynous), male gibbons (who form single pair-bonds with females), and saddleback tamarin monkeys (in which a female keeps two males).

"Even in the most social mammals . . . there is no evidence that males even recognize their own offspring."

Diamond, J. (1997). *Why Is Sex Fun?* New York: Basic Books.

"In humans . . . the dominant mode of relationships is of monogamy . . ."

Diamond, J. (1997). *Why Is Sex Fun?* New York: Basic Books.

p. 243 **"The human cochlea is so sensitive that it can detect vibration as small as the diameter of an atom (0.3 nm) and it can resolve time intervals down to 10μs . . ."**

Hudspeth, A. J. (1997). How hearing happens. *Neuron* 19: 947–950.

p. 244 **". . . some animals employ systems that are exotic compared to ours."**

Hughes, H. C. (1999). *Sensory Exotica: A World Beyond Human Experience.* Cambridge, MA: MIT Press.

"Sharks have an *electrical* sense . . ."

While snorkeling in the Caribbean last spring, I actually heard the electrical discharges of tropical fish for myself, which sounded like a high-pitched, rapid clicking sound. With human ears, I could hear but not localize the source; the sharks' sense would be distinct from hearing and permit them to locate their prey.

p. 245 **"The basic function and structure of genes is also common to all animals . . ."**

See: Colamarino, S., and M. Tessier-Lavigne. (1995). The role of the floorplate in axon guidance. *Annual Review of Neuroscience* 18: 497–529. Deacon, T. W. (1997). What makes the human brain different? *Annual Review of Anthropology* 26: 337–357.

Friedman, G., and D. D. O'Leary. (1996). Retroviral misexpression of engrailed genes in the chick optic tectum perturbs the topographic targeting of retinal axons. *Journal of Neuroscience* 16(17): 5498–5509.

Kennedy, T. E., T. Serafini, J. R. de la Torre, and M. Tessier-Lavigne. (1994). Netrins are diffusible chemotropic factors for commissural axons in the embryonic spinal chord. *Cell* 78: 425–435.

". . . Balaban removed the auditory cortices from Japanese quail embryos and surgically implanted them into the brains of embryonic chickens . . ."

Balaban, E., M. A. Teillet, and N. Le Douarin. (1988). Application of the quail-chick chimera system to the study of brain development and behavior. *Science* 241(4871): 1339–1342.

p. 248 **"During a typical day, chimpanzees . . . associate in temporary parties that vary in size and membership, much as humans do."**

Ujhelyi, M. (1996). Is there any intermediate stage between animal communication and language? *Journal of Theoretical Biology* 180(1): 71–76.

". . . biologists have found an inverse relationship between brain size and digestive tract size . . ."

Allman, J. M. (1999). *Evolving Brains.* New York: Scientific American Library/W.H. Freeman.

p. 249 **"The total amount of energy available to an organism is limited, forcing an evolutionary trade-off between brain size and digestive tract size."**

Aiello, L, and P. Wheeler. (1995). The expensive tissue hypothesis: The brain and the digestive system in human and primate evolution. *Current Anthropology* 36: 199–221.

"Geneticists found that humans have lost an ability . . . to create vitamin C internally . . ."

Ha, M.N., F. L. Graham, C. K. D'Souza, W. J. Muller, S. A. Igdoura, and H. E. Schellhorn. (2004). Functional rescue of vitamin C synthesis deficiency in human cells using adenoviral-based expression of murine l-gulono-gamma-lactone oxidase. *Genomics* 83(3): 482–492.

Stone, I. (1979). Homo sapiens ascorbicus, a biochemically corrected robust human mutant. *Medical Hypotheses* 5(6): 711–721.

". . . big brains carry with them a high biological price tag . . ."

This passage borrows liberally from: Allman, J. M. (1999). *Evolving Brains.* New York: Scientific American Library/W.H. Freeman, p. 160.

p. 250 **"Well-connected female baboons have more babies who receive better care . . ."**

Zimmer, C. (March 4, 2008). Sociable, and smart. *The New York Times,* pp. D1, D4.

" 'Spotted hyenas live in a society just as large and complex as a baboons.' "

Zimmer, C. (March 4, 2008). Sociable, and smart. *The New York Times,* p. D1.

". . . similar evolutionary forces worked independently to arrive at a similar adaptive solution . . ."

Holekamp, K. (2006). Spotted hyenas. *Current Biology* 16: R944–R945.

p. 251 **". . . tool *making* as opposed to mere tool *use* . . . represented a major cognitive leap . . ."**

Tattersall, I. (January 2000). Once we were not alone. *Scientific American* 282(1): 57–62.

"Stone tools are thus the first evidence we have of the birth of *symbolic thought* . . ."

Tattersall, I. (January 2000). Once we were not alone. *Scientific American* 282(1): 57–62.

". . . humans must have brought musical instruments with them when they left Africa for Europe."

Cross, I. (2006). The origins of music: Some stipulations on theory. *Music Perception* 24(1): 79–82.

p. 252 **". . . workmanship [their stone tools] that displayed . . . 'an exquisite sensitivity to the properties' of these materials."**

Tattersall, I. (January 2000). Once we were not alone. *Scientific American* 282(1): 61.

" 'Clearly . . . these people were *us*.' "

Tattersall, I. (January 2000). Once we were not alone. *Scientific American* 282(1): 61.

p. 257 **". . . (my dog Shadow distinguishes several different toys by name including his 'fuzzy man' from his 'Cat in the Hat') . . ."**

My dog Isabella before him could pick out ten different items by name, including *newspaper, ball, Frisbee, bed,* and *bone.* See also:

Kaminski, J., J. Call, and J. Fisher. (2004). Word-learning in a domestic dog: Evidence for fast mapping. *Science* 304: 1682–1683.

". . . but this only demonstrates their [animals'] ability to *link* a visual or acoustic stimulus with an object."

As a well-known example, Ivan Pavlov showed that dogs can learn to associate the sound of a bell with the presentation of food. My dog associates the word *cookie* with the treats I keep in the pantry. But there are important differences between associations and a formal concept of naming, the awareness that the name and the object it refers to are two separate things. I can talk about cookies without you expecting to get one; I cannot do that with my dog, as clever as he is. That's one of the differences between a "name" and an "association."

p. 258 **". . . every day humans produce or hear utterances that have never before been spoken, and yet we understand them."**

I'm glossing over many important details associated with the expandable property of language. These are dealt with in Steven Pinker's *The Language Instinct* and in the article by Hauser, Chomsky, and Fitch (see below for full references).

One important concept in the discussion is the capacity for *recursion,* a cognitive operation believed by many to be unique to humans and central to language. Briefly, recursion describes the formal way in which expressions can be expanded indefinitely. It can be thought of as a set of instructions that can loop around back to the beginning, or in computer science jargon, a "routine that can call itself." Take for example the following instructions for how to wash a dirty soup pot:

Routine for Washing a Dirty Soup Pot

1. Rinse with water.
2. Add soap.
3. Scrub with brush, sponge, or scouring pad until it appears to be clean.
4. Rinse.
5. Check to see if it is clean. If yes, go to step 6. If not, execute the instructions "Routine for Washing a Dirty Soup Pot."

6. Dry.
7. Stop (you're done).

The "branching loop" in step 5 is what makes this recursive. It is a routine that can expand indefinitely. Human languages have sentences that can do the same thing of course, as illustrated in the main text.

The notion that recursion is central to human language has been challenged by Daniel Everett. The very fact that there is a debate between the Chomskians and Everett serves to support my point that there is *not* a single unique element that humans possess that gives us language; rather, animal and human communications form a continuum and many elemental operations show up along the continuum. The fact that at least one human group lacks this operation (recursion) makes the claim untenable that it is both unique to humans and necessary for human language.

For the standard view on what constitutes language, see for example: Pinker, S. (1994). *The Language Instinct.* New York: Morrow. Or see: Hauser, M. D., N. Chomsky, and W. T. Fitch. (2002). The faculty of language: What is it, who has it and how did it evolve? *Science* 298: 1569–1579.

For a dissenting view, see: Everett, D. L. (2005). Cultural constraints on grammar and cognition in Pirahã. *Current Anthropology* 46(4): 621–646.

p. 261 **"'Animals of many kinds are social . . .'"**

Darwin, C. (1981). *The Descent of Man and Selection in Relation to Sex.* Princeton, NJ: Princeton University Press, pp. 161–163. (Original work published 1871.)

"'In order that primeval men, or the apelike progenitors of man, should become social . . .'"

Darwin, C. (1981). *The Descent of Man and Selection in Relation to Sex.* Princeton, NJ: Princeton University Press, pp. 161–163. (Original work published 1871.)

p. 262 **"In one experiment, she [Haselton] asked people to think about how much they love their partner and then try to suppress thoughts of other people they find sexually attractive."**

This description taken from Zimmer, C. (January 17, 2008). Romance is an illusion [Electronic version]. *Time.* Retrieved March 10, 2008, from http://www.time.com/time/magazine/article/0,9171,1704665,00.html.

p. 264 **". . . European starlings can *learn* syntactic recursion."**

Gentner, T. Q., K. M. Fenn, D. Margoliash, and H. C. Nusbaum. (2006). Recursive syntactic pattern learning by songbirds. *Nature* 440: 1204–1207.

". . . white-crowned sparrows can assemble an entire song in proper sequence when exposed to only fragments of that song . . ."

Rose, G. J., F. Goller, H. J. Gritton, S. L. Plamondon, A. T. Baugh, and B. G. Cooper. (2004). Species-typical songs in white-crowned sparrows tutored with only phrase pairs. *Nature* 432: 753–758.

p. 266 **"It is important, when considering animal music, to distinguish between musical expression and musical experience."**

Jerison, H. (1999). Paleoneurology and the biology of music. In *The Origins of Music,* edited by N. L. Wallin, B. Merker, and S. Brown. Cambridge, MA: MIT Press, pp. 177–196.

"Evolution endowed the musical brain with a perception-production link that most mammals lack . . . We *hear* music, then *sing* it."

Merker, B. (2006). The uneven interface between culture and biology in human music. *Music Perception* 24(1): 95–98.

p. 267 **". . . most [mammals] do not have the capacity to imitate a sound they've heard . . ."**

Merker, B. (2006). The uneven interface between culture and biology in human music. *Music Perception* 24(1): 95–98.

"This vocal learning ability is believed to have come from an evolutionary modification to the basal ganglia . . ."

Patel, A. D. (2006). Musical rhythm, linguistic rhythm, and human evolution. *Music Perception* 24(1): 99–104.

". . . Brodmann area 44 . . . "

Iacoboni, M., I. Molnar-Szakacs, V. Gallese, G. Buccino, J. C. Mazziotta, and G. Rizzolatti. (2005). Grasping the intentions of others with one's own mirror neuron system. *Public Library of Science Biology* 3(1): e79.

". . . FOXP2 gene . . ."

Wade, N. (October 18, 2007). Neanderthals may have had gene for speech [Electronic version]. *The New York Times*. Retrieved March 10, 2008, from http://www.nytimes.com/2007/10/18/science/19speech.html?partner=rssnyt&emc=rss.

". . . a genetic variant in microcephalin . . ."

Gazzaniga, M. S. (2007). Are Human Brains Unique? From John Brockmans' *Edge*, April 10, 2007. www.edge.org.

p. 269 **". . . there are no immaterial, vitalistic, or supernatural processes involved in creating the experience we call consciousness . . ."**

Bunge, M. (1980). *The Mind-Body Problem: A Psychological Approach.* New York: Pergamon.

p. 270 **". . . spontaneous intelligence."**

Johnson, S. (2001). *Emergence: The Connected Lives of Ants, Brains, Cities, and Software.* New York: Scribner.

"As essayist Adam Gopnik says . . ."

Gopnik, A. (2006). Death of a Fish. From *Through the Children's Gate.* New York: Knopf, p. 258.

". . . human music functions as an *honest signal* . . ."

There exists controversy on this point. Cognitive psychologist Jamshed Bharucha points out that a skilled performer can effectively evoke an emotion that he doesn't feel. When a performer plays a sad piece, it's not necessarily because he wishes to communicate his sadness. The piece may simply be on the program for that night, and the performer knows how to perform it to elicit sadness. David Byrne corroborated this in an interview with me—he doesn't always *feel* sad when he's singing a sad song, but he's learned what are essentially a set of tricks or devices to evoke sadness or other emotions required for the emotional delivery of the song.

Bharucha asks us to consider an evolutionary context: A man courting a woman by expressing his love to her. He can deceive her using language, i.e., convince her that he loves her even if he doesn't, just so he can get sex. The same would be true of music. He could appear to pour his heart out to her if he is a skillful musician—even if he doesn't feel undying love—and thereby gain access to sex. Are language and music different here? Is music an honest signal?

Music may have started out as an honest signal—something difficult to fake. But a sort of arms race may have developed. Some humans would have learned how to fake the emotions in music. Through intensive training, for example, they might learn to *appear* sad or in love or happy even when they are not. Actors do the same thing with language, of course. Actors essentially lie for a living. To be successful, they have to make you think that they are someone whom they are not, and that the words they are speaking are being uttered spontaneously and on their own, even when (most of the time) those words were written by someone else ahead of time.

If we accept the honest signal hypothesis, it doesn't have to mean that music is still a foolproof honest signal, only that it once was (and perhaps still is) a *more* honest signal than language. We can speculate about why music might be better at this: Because of music's structure and internal complexity, music typically packs much more information into a phrase than language does. This might make it more difficult to fake honesty because so many more dimensions of expression would need to be manipulated than simply the words and linguistic prosody.

It's worth noting that once expert singers learned to fool ordinary listeners, there would be increased evolutionary pressure for listeners to become more discriminating, which would lead to pressure for the singers to become more skilled. If music started out as an honest signal, with connections to all the right emotional and motivation centers in the human brain, these more (evolutionarily) recent developments may not have yet effected commensurate changes in neural wiring, or the changes may be still under way. This could account for why skillful musicians can move us to laughter and to tears: Our cognitive appraisal system *knows* that we are being "lied to," and yet all the emotional buttons are still being pushed. The result is a deeply emotional reaction that is bound up with an aesthetic and cognitive appreciation for what is going on.

p. 271 **". . . love is like jumping off a cliff."**

Tennant, A. (director), J. Lassiter, W. Smith, T. Zee (Producers), and K. Bisch (writer). (2005). *Hitch* [Motion picture]. United States: Columbia Pictures.

p. 274 **". . . the Romeo-and-Juliet love (I'd kill myself for this person) . . ."**
Blue Öyster Cult's "(Don't Fear) the Reaper" is perhaps the first teenage suicide pact song in rock music.
Roeser, D. (1976). (Don't fear) the reaper [Recorded by Blue Öyster Cult]. On *Agents of Fortune* [45rpm record]. Columbia Records.

p. 283 **". . . I've intentionally avoided becoming distracted by questions such as 'What are the greatest/most popular songs of all time?' or . . . 'What are the most *influential* songs of all time?' "**
The RIAA, a recording industry lobbying group, along with the National Endowment for the Arts, sponsored a Greatest Songs of the Twentieth Century project in 2001, and voted "Over the Rainbow" as the top song, followed by "White Christmas." Such lists are not just subjective, but can yield bizarre results. Is "Fight for Your Right (To Party)" by the Beastie Boys (#191) four songs better than Cole Porter's "Night and Day" (#195)? What kind of list puts "Take Me Out to the Ball Game" (#8) above "You've Lost That Lovin' Feelin' " (#9). And how did "Achy Breaky Heart" (#258) beat out "All Along the Watchtower" (#365) and "How High the Moon" (#317)?
Arlen, H., and Harburg. E. Y. (1939). Over the rainbow [Recorded by Judy Garland]. On *Over the Rainbow* [LP]. Pickwick Records.
Berlin, I. (1940). White Christmas [Recorded by Bing Crosby and Marjorie Reynolds]. On *Holiday Inn* [LP]. (1942).
Beastie Boys. (1986). (You gotta) Fight for your right (to party!). On *Licensed to Ill* [CD]. Def Jam Records.
Dylan, B. (1967). All along the watchtower. On *John Wesley Harding* [LP]. Nashville, TN: Columbia Records.
Hamilton, N., and Lewis. M. and (1940). How high the moon [Recorded by Benny Goodman and His Orchestra]. On *How High the Moon* [45rpm record]. Columbia Records.
Porter, C. (1932). Night and day [Recorded by Fred Astaire]. On *Night and Day: Fred Astaire: Complete recordings Vol. 2 1931–1933* [CD]. Naxos Nostalgia. (2001).
Norworth, J. (1908). Take me out to the ball game [Recorded by Harry MacDonough]. On *Take Me Out to the Ball Game* [Wax cylinder]. Victor Records.
Spector, P., B. Mann, and C. Weil. (1965). You've lost that lovin' feelin'

[Recorded by the Righteous Brothers]. On *You've Lost That Lovin' Feelin'* [45rpm record]. Philles Records.

Von Tress, D. (1992). Achy breaky heart [Recorded by Billy Ray Cyrus]. On *Some Gave All* [CD]. Mercury Records.

Grateful acknowledgment is made to the following for permission to reprint:

Pages 1–2. “Homegrown Tomatoes.” Words and music by Guy Clark. Copyright © 1983 by EMI April Music Inc. and GSC Music. All rights controlled and administered by EMI April Music Inc. All rights reserved. International copyright secured. Used by permission.

Page 5. “O Kazakhstan.” From *Borat: Cultural Learnings of America for Make Benefit Glorious Nation of Kazakhstan*. Words and music by Sacha Baron Cohen, Erran Baron Cohen, Peter Baynham, Daniel Greaney, Anthony Hines, Patton Oswalt, Todd Phillips and Jeffrey Poliquin. Copyright © 2006 by T C F Music Publishing, Inc. and Fox Film Music Corp. All rights reserved. Used by permission.

Page 5. “Dirt Bike.” Words and music by John Linnell and John Flansburgh. Copyright © 1994 by TMBG Music. All rights on behalf of TMBG Music administered by Warner-Tamerlane Publishing Corp. All rights reserved. Used by permission of Alfred Publishing Co., Inc.

Pages 72–73. “If I Had a Rocket Launcher.” Written by Bruce Cockburn. Copyright © 1984 by Golden Mountain Music Corp. (SOCAN). Used by permission.

Page 87. “Log Blues.” From *The Ren & Stimpy Show*. Words and music by Charlie Brissette and Christopher Reccardi. Copyright © 1993 by Nickelodeon, Inc. All rights administered by Sony/ATV Music Publishing LLC, 8 Music Square West, Nashville, TN 37203. International copyright secured. All rights reserved.

Page 128. “At Seventeen.” Words and music by Janis Ian. Copyright © 1975 (renewed 2003) by Mine Music Ltd. All rights for the U.S.A. and Canada controlled and administered by EMI April Music Inc. All rights reserved. International copyright secured. Used by permission.

Page 129. “Death Is Not the End.” Written by Bob Dylan. Copyright © 1988 by Special Rider Music. All rights reserved. Used by permission.

Pages 130–31. “The Revolution.” Words and music by David Byrne. Copyright © by Moldy Fig Music, Inc. Used by permission.

Pages 186–87. “Natural Science.” By Neil Peart. Music by Geddy Lee and Alex Lifeson. Copyright © 1980 by Core Music Publishing. All rights reserved. Used by permission of Alfred Publishing Co., Inc.

Page 230. “Happy Together.” Words and music by Garry Bonner and Alan Gordon. Copyright © 1966, 1967 by Alley Music Corp. and Trio Music Company. Copyright renewed. International copyright secured. All rights reserved. Used by permission.

Page 233. “Cupid’s Got a Brand New Gun.” Words and music by Michael Penn. Copyright © by 1989 Bucket Brigade Songs (BMI). All rights reserved. Used by permission.

Page 235. “Paralyzed.” Written by Rosanne Cash. Copyright © 1990 by Chelcait Music (BMI), administered by Bug. All rights reserved. Used by permission.

Page 275. “I’m Not in Love.” Words and music by Eric Stewart and Graham Gouldman. Copyright © 1975 (renewed 2003) by Man-Ken Music Ltd. All rights controlled and administered by EMI Blackwood Music Inc. All rights reserved. International copyright secured. Used by permission.

Pages 285–87. “The Randall Knife.” Words and music by Guy Clark. Copyright © 1983 by EMI April Music Inc. and GSC Music. All rights controlled and administered by EMI April Music Inc. All rights reserved. International copyright secured. Used by permission.

Acknowledgments

I'd like to thank the musicians and academics who generously let me interview them for this book: Jonathan Berger, Michael Brook, David Byrne, Ian Cross, Rodney Crowell, Don DeVito, Jim Ferguson, David Huron, Joni Mitchell, Sandy Pearlman, Oliver Sacks, Pete Seeger, and Sting. I am grateful to McGill University for providing me with a stimulating and supportive environment in which to work. My editor, Stephen Morrow at Dutton, was indispensable in bringing this book to reality. It has been a joy and a comfort to work with him, and he has contributed greatly, from the initial concept for *The World in Six Songs* (which was his) through every stage of the writing and editing. My superb agent, Sarah Chalfant, along with Edward Orloff and everyone else at the Wylie Agency, provided guidance and support throughout. Thanks to Duttoners Erika Imranyi, Christine Escalante, and Susan Schwartz for taking up the slack on details too numerous to mention, and to Lisa Johnson, Beth Parker, Andy Heidel, Sarah Muszynski, Marie Coolman, and Mary Pomponio for helping my work reach a wider audience. To Kathy Schenker, Tracy Bufferd, Dave Whitehead, and Michael Hausman: Thanks for your unflappable way of making difficult things easy.

My students read drafts of this book and provided helpful advice: Vanessa Park-Thompson, Mike Rud, and Anna Tirovolas. Bianca Levy performed tireless and rigorous background research on both the science and the music sides, making many helpful and insightful suggestions. My girlfriend was a tremendous source of emotional support and gave generously of her time to listen to and comment on successive drafts of the manuscript, making it immeasurably better. This book would not be what it is without her. Thanks also to the following for their thoughtful reading of the manuscript and helpful comments: Professors Jamshed Bharucha (Provost and Department of Psychology, Tufts University), Dennis Drayna (National Institutes of Health), Charles Gale (Department of Physics, McGill University), Frederic Guichard (Department of Biology, McGill University), David Huron (Department of Music, Ohio State University), Jeff Mogil (Department of Psychology, McGill University), Monique Morgan (Department of English, McGill University), Frank Russo (Department of Psychology, Ryerson University), Barbara Sherwin (Department of Psychology, McGill University), Wilfred Stone (Department of English, Stanford University), and my friends Len Blum, Parthenon Huxley, and Jeff Kimball. In all that I've done over the past twenty years, I've found great inspiration from Lew Goldberg (Oregon Research Institute), a rigorous scientist, challenging mentor, and dear friend. I've also benefited enormously from reading and interacting with Oliver Sacks, Daniel Dennett, Roger Shepard, Michael Posner, David Huron, and Ian Cross. It is by standing on the shoulders of these giants that I have been shown things I didn't know existed.

Index

About the Author

Daniel J. Levitin is the James McGill Professor of Psychology, Neuroscience, and Music at McGill University, where he runs the Laboratory for the Study of Music Perception, Cognition, and Expertise. He is the author of the *New York Times* bestseller *This Is Your Brain on Music: The Science of a Human Obsession,* a finalist for the *Los Angeles Times* Book Prizes, the Quill Award (best debut author), and named as one of the *Globe & Mail*'s Top Books of 2006. He lives in Montreal, Canada.